Gene and Cell Therapies

Pharmaceuticals, Health Economics and Market Access

Editor:
Mondher Toumi

Gene and Cell Therapies: *Market Access and Funding*
Eve Hanna and Mondher Toumi

For more information about this series, please visit: https://www.crcpress.com/Pharmaceuticals-Health-Economics-and-Market-Access/book-series/CRCPHEMA

Gene and Cell Therapies

Market Access and Funding

Eve Hanna
Mondher Toumi

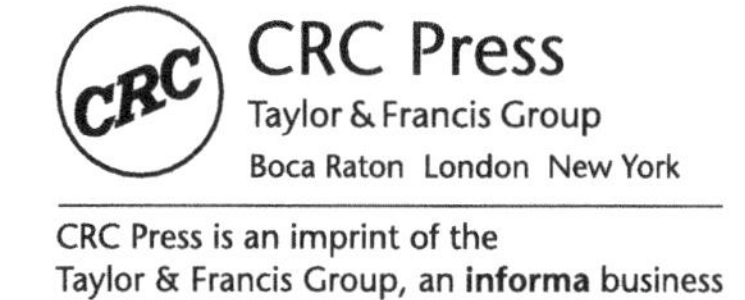

CRC Press is an imprint of the
Taylor & Francis Group, an **informa** business

First edition published 2020
by CRC Press
6000 Broken Sound Parkway NW, Suite 300, Boca Raton, FL 33487-2742

and by CRC Press
2 Park Square, Milton Park, Abingdon, Oxon, OX14 4RN

CRC Press is an imprint of Taylor & Francis Group, LLC

ISBN: 978-0-367-48384-5 (hbk)
ISBN: 978-0-367-40809-1 (pbk)
ISBN: 978-0-367-80920-1 (ebk)

Typeset in Palatino
by Lumina Datamatics Limited

Contents

Series preface xiii
Foreword xv
List of abbreviations xvii
Authors xxi

Chapter 1 Introduction to cell and gene therapies concepts and definitions in the US and the EU 1
1.1 Introduction to EU regulation (EC) No 1394/2007 of ATMPs 2
1.1.1 Advanced therapy medicinal products definitions 3
1.1.1.1 Gene therapy medicinal product 3
1.1.1.2 Somatic cell therapy medicinal product and tissue-engineered product 3
1.1.1.3 Combined therapy medicinal product 5
1.1.1.4 Classification by donor 5
1.1.2 Borderline classification 5
1.1.3 Hospital exemption 6
1.1.4 Transitional period 7
1.1.4.1 Number of ATMPs before the regulation 7
1.1.4.2 Transitional period timelines 7
1.2 ATMPs approved in EU 7
1.2.1 Chondrocelect® 8
1.2.2 Glybera® 8
1.2.3 MACI® 8
1.2.4 Provenge® 8
1.2.5 Holoclar® 9
1.2.6 Imlygic® 9
1.2.7 Strimvelis® 9
1.2.8 Zalmoxis® 9
1.2.9 Spherox® 10
1.2.10 Alofisel® 10
1.2.11 Yescarta® 10

1.2.12 Kymriah® ... 11
1.2.13 Luxturna® ... 11
1.2.14 Zynteglo® ... 11
1.3 Introduction to regenerative medicine advanced therapy (RMAT) designation ... 11
1.4 Approved cell and gene therapies in the US ... 15
1.5 Conclusion ... 16
References ... 16

Chapter 2 Cell and gene therapies: genuine products and potential for dramatic value ... 19
2.1 Cell and gene therapies specificities ... 19
2.1.1 Personalized therapies and manufacturing specificities ... 19
2.1.2 "Once and done" concept ... 20
2.1.3 Curative or transformative effect with downstream long-term outcomes ... 20
2.1.4 High up-front cost ... 22
2.2 Cell and gene therapies value ... 23
2.2.1 Value for patients ... 23
2.2.1.1 Curing diseases and saving patients from death ... 24
2.2.1.2 Survival prolongation ... 24
2.2.1.3 Halting disease progression ... 25
2.2.1.4 Improving health-related quality of life (HRQoL) ... 25
2.2.2 Value for the society ... 25
2.2.2.1 Cost saving ... 25
2.2.2.2 Decrease the burden of disease ... 26
2.2.2.3 Changes in medical paradigm ... 26
2.2.2.4 Promising prospect ... 27
2.2.2.5 Regenerative medicine could slow the aging process ... 28
2.3 Advanced therapies: a heterogeneous class ... 28
2.3.1 Value drivers ... 28
2.3.2 Different archetypes ... 28
2.4 Conclusion ... 29
References ... 29

Chapter 3 Cell and gene therapies: regulatory aspects in the US and the EU ... 31
3.1 Regenerative medicine regulatory pathway in EU ... 31
3.1.1 ATMP specific committee: committee for advanced therapies (CAT) ... 31
3.1.1.1 Members ... 31
3.1.1.2 CAT scientific recommendation on advanced therapy classification ... 32

3.1.2 Premarketing authorization ... 32
3.1.2.1 Risk-based approach ... 32
3.1.2.2 Manufacturing quality requirements ... 34
3.1.2.3 Nonclinical data ... 34
3.1.2.4 Clinical data ... 34
3.1.3 Marketing authorization ... 35
3.1.3.1 ATMPs guidelines ... 35
3.1.3.2 Marketing authorization application ... 36
3.1.3.3 Centralized marketing authorization procedure ... 38
3.1.3.4 Marketing authorization timelines ... 39
3.1.3.5 Marketing authorization fees ... 39
3.1.4 Regulatory early access tools and development support ... 41
3.1.4.1 Accelerated assessment ... 41
3.1.4.2 Conditional MA ... 42
3.1.4.3 CHMP Compassionate use opinion ... 42
3.1.4.4 Development support: PRIME ... 42
3.1.4.5 Innovation task force (ITF) ... 48
3.1.4.6 SME office ... 48
3.1.4.7 Scientific advice and protocol assistance ... 48
3.1.5 Post-authorization requirements ... 49
3.1.6 Incentives of the ATMPs regulation ... 50
3.2 Regenerative medicine regulatory pathway in the US ... 51
3.2.1 FDA expedited programs ... 51
3.2.1.1 Fast-track designation ... 51
3.2.1.2 Breakthrough therapy designation ... 52
3.2.1.3 Regenerative medicine advanced therapy designation ... 52
3.2.1.4 Priority review designation ... 53
3.2.1.5 Accelerated approval ... 53
3.2.2 Application for RMAT designation ... 54
3.2.3 Post-approval requirements ... 54
3.2.4 Interactions between sponsors and CBER review staff ... 55
3.3 Conclusion ... 55
References ... 57

Chapter 4 The need for new HTA reference case for cell and gene therapies ... 59
4.1 EU Reimbursement: current HTA frameworks in Europe ... 59
4.2 Assessment of current framework applicability to ATMPs in England: NICE mock appraisal ... 61
4.2.1 Hypothesis: benefits and costs ... 61
4.2.1.1 Summary of results ... 62

4.3 Review of HTA experience in Europe 63
4.3.1 France 63
4.3.1.1 Glybera® 63
4.3.1.2 Chondrocelect® 66
4.3.1.3 Holoclar® 67
4.3.1.4 Zalmoxis® 67
4.3.1.5 Yescarta® 67
4.3.1.6 Kymriah® 67
4.3.1.7 Luxturna® 68
4.3.2 Germany 68
4.3.2.1 Glybera® 68
4.3.2.2 Provenge® 69
4.3.2.3 Imlygic® 69
4.3.2.4 Zalmoxis® 70
4.3.2.5 Alofisel® 70
4.3.2.6 Kymriah® 70
4.3.2.7 Yescarta® 72
4.3.2.8 Luxturna® 72
4.3.3 England 73
4.3.3.1 Provenge® 73
4.3.3.2 Imlygic® 74
4.3.3.3 Holoclar® 75
4.3.3.4 Strimvelis® 75
4.3.3.5 Kymriah® 75
4.3.3.6 Yescarta® 76
4.3.3.7 Luxturna® 76
4.3.3.8 Alofisel® 76
4.3.3.9 Spherox® 77
4.3.4 Scotland 77
4.3.4.1 Yescarta® 77
4.3.4.2 Kymriah®-DLBCL 77
4.3.4.3 Kymriah®-B cell ALL 78
4.3.5 Italy 78
4.3.6 Spain 78
4.4 US reimbursement 78
4.4.1 ICER initiative: methods adaptation in the assessment of potential cures 79
4.4.2 Advanced therapies reimbursement in the US 80
4.4.3 Review of CAR-T cell therapies case study: Kymriah® and Yescarta® 80
4.4.3.1 ICER assessment of CAR-T therapies 80
4.4.3.2 CMS coverage 80

4.5 Challenges at HTA level ... 81
4.5.1 Uncertainty ... 81
4.5.2 Uncertainty around short-term efficacy ... 81
4.5.2.1 No comparative trials ... 81
4.5.3 Uncertainty around long-term efficacy and safety ... 82
4.5.4 Uncertainty around cost-effectiveness ... 82
4.5.5 HTA agencies lack of flexibility and different requirements from regulators ... 83
4.5.6 Conclusion on the valuation challenges ... 84
4.6 Commercial challenges ... 84
4.7 The consequences of market access delays ... 84
4.7.1 Case study: Provenge® ... 86
4.8 Conclusion ... 86
References ... 87

Chapter 5 How to mitigate uncertainties and HTA risk-averse attitude? ... 91
5.1 Conditional reimbursement ... 91
5.1.1 Limitation ... 92
5.1.2 The use of conditional reimbursement ... 92
5.1.3 Conditional reimbursement for gene therapies ... 93
5.2 The use of registries to collect long-term, real-world data ... 94
5.2.1 EMA guidelines on establishing CAR-T patient registries ... 94
5.3 Improving the acceptability of single-arm with historical control trial design ... 94
5.4 Recommendations of adaptation of economic evaluation ... 95
5.4.1 ICER initiative ... 95
5.4.2 Our suggestion for adaptation of economic modeling requirements ... 96
5.5 Early dialogue with regulators and payers ... 96
5.5.1 Limitation ... 96
5.5.2 Benefit-risk assessment for HTA early advice ... 96
5.6 Conclusion ... 99
References ... 100

Chapter 6 Cell and gene therapies funding: challenges and solutions for patients' access ... 101
6.1 Current healthcare spending ... 101
6.2 Current cost-containment policies ... 102
6.2.1 External reference pricing ... 103
6.2.2 Internal reference pricing ... 103

6.2.3 Value-based pricing 103
6.2.4 Cost-based pricing 103
6.2.5 Profit control as an indirect price control 104
6.2.6 Discounts, rebates, expenditure caps, and price-volume agreements 104
6.3 Proposed funding models for high-cost therapies 104
6.3.1 Financial agreements 105
6.3.1.1 Bundle payment, episode of care 105
6.3.1.2 Rebates and discounts 106
6.3.1.3 Price caps/volume caps per patients or per target population 107
6.3.1.4 Price-volume agreements 107
6.3.1.5 Healthcare loans/credits 108
6.3.1.6 Fund-based payment divided in three subcategories: pooled funding, national silo funds, and special international fundraising 108
6.3.1.7 Intellectual-based payment 109
6.3.2 Health outcomes-based agreements 110
6.3.2.1 Payment-for-performance 110
6.3.2.2 Annuity payment 115
6.3.2.3 Coverage with evidence development 115
6.3.3 Healthcoin 116
6.4 Comparison of the proposed funding models features 116
6.5 Examples of adoption of new payment models 119
6.5.1 Suggestion of a new pricing scheme in Germany 119
6.5.2 Germany is piloting a pay for performance scheme 119
6.5.3 Pilot test for annuity payment in the US 119
6.6 Review of the successful examples of CAR-T cells 119
6.6.1 France 120
6.6.2 England 120
6.7 Future trends: which funding model could be adopted for advanced therapies? 120
6.7.1 A potentially sustainable short-term solution: advanced therapies fund 121
References 122

Chapter 7 Conclusion 127
7.1 The long-awaited cell and gene therapies era is finally becoming reality 127
7.2 Regulators prepared for the phenomena 133
7.3 HTA agencies are being conservative, not willing to adapt the current decision frameworks 134

7.4 Funding remains a serious challenge and little is done today......... 134
7.5 Cell and gene therapies challenge the state-of-the-art business model of pharma industry.. 135
7.6 Conclusion.. 137
References.. 138

Index 141

Series preface

The major advances in the field of biotechnology and molecular biology in the twenty-first century have led to a better understanding of the pathophysiology of diseases. A new generation of biopharmaceuticals has emerged including a wide and heterogeneous range of innovative therapies. It aims to prevent or treat chronic or serious life-threatening diseases, which were previously considered as incurable. This unique book series focuses on how the regulatory environment has evolved to analyze and review those therapies while HTA agencies and payers remain resistant. It provides an insight into current learning by demonstrating how those products will be accessible and which policy changes will be required to permit patient access.

Foreword

The advent of cell and gene therapy is likely to be the defining healthcare event of this decade. After years of promise, new therapies are becoming available. In this book, Eve Hanna and Mondher Toumi note that by September 2019, around 47 cell and gene therapies have been approved globally and many more are in the pipeline. Not only will this revolutionize healthcare itself, it is likely to bring about major changes in how we regulate and pay for new health technologies.

This book is ideal for someone wanting to get up to speed with all the issues surrounding these new therapies. It is very comprehensive, starting with the key concepts surrounding cell and gene therapies and the various definitions used in the United States and the European Union. For example, it is now considered better to call these therapies "transformative" rather than "curative." The authors also discuss the fact that although the therapies have specific characteristics, typically a single dose with long-lasting effects, they also share characteristics with other important groups of medicines, such as those for rare diseases.

Building on a discussion of the characteristics of cell and gene therapies and their potential value, the authors go on to discuss the challenges they raise for regulation (by drug licensing authorities), reimbursement approval (by health technology assessment agencies), and price negotiations (with healthcare payers). It is stressed that the change in the medical paradigm brought about by these therapies has important implications for the types of clinical evidence considered and the projections of long-term outcomes. The "up-front" nature of the costs and the uncertainty around value in the long term have implications for both how we assess these therapies and how we pay for them. Chapters 3–6 discuss these issues in detail, giving lots of interesting examples.

The concluding chapter discusses ways forward and provides some excellent insights concerning what needs to change if society is going to harness the full benefits offered by these therapies. This book is timely and makes an important contribution to our understanding of the potential revolution in healthcare that we face at this interesting time.

Michael Drummond
Professor of Health Economics
University of York, UK

List of abbreviations

AADC Aromatic L-Amino Acid Decarboxylase
AAV Adenovirus Associated Viral Vector
AB Actual Benefit *(Service Médical Rendu)*
ACO Affordable Care Organizations
ADA Adenosine Deaminase Deficiency
AIFA Italian Medicines Agency *(Agenzia Italiana del Farmaco)*
ALD Adrenoleukodystrophy
ALL Acute Lymphoblastic Leukaemia
ARM Alliance for Regenerative Medicines
ASCT Autologous Stem Cell Transplant
ATMP Advanced Therapy Medicinal Products
ATU Temporary authorisation for use *(Autorisation Temporaire d'Utilisation)*
AWU Annual Work Unit
BCMA B-Cell Maturation Antigen
BLA Biologics License Application
CADTH Canadian Agency for Drugs and Technologies
CAR Chimeric Antigen Receptor
CAT Committee for Advanced Therapies
CATP Combined Therapy Medicinal Product
CBER Centre for Biologics Evaluation and Research
CDF Cancer Drugs Fund
CED Coverage with Evidence Development
CEPS *Comité Economique des Produits de Santé*
CFR Code of Federal Regulations
CHMP Committee for Medicinal Products for Human Use
CI Confidence Interval
CMS Centers for Medicare & Medicaid Services
COMP Committee on Orphan Medicinal Products
CR Complete Remission
CRISPR Clustered Regularly Interspaced Short Palindromic Repeats
CRPC Castrate Resistant Prostate Cancer
CTD Common Technical Document

CTMP	Cell Therapy Medicinal Product
DLBCL	Diffuse Large B-Cell Lymphoma
DMD	Duchenne Muscular Dystrophy
DNA	Deoxyribonucleic Acid
DRG	Diagnosis Related Group
EC	European Commission
EMA	European Medicines Agency
ERP	External Reference Pricing
EU	European Union
FDA	Food and Drug Administration
GCP	Good Clinical Practice
GDP	Gross Domestic Product
GMP	Good Manufacturing Practices
GTMP	Gene Therapy Medicinal Product
GVHD	Graft Versus Host Disease
HAS	*Haute Autorité de Santé*
HCT/P	Human Cells, Tissues, or Cellular or Tissue-Based Products
HHS	Health and Human Services
HLA	Human Leukocyte Antigen
HPC	Hematopoietic Progenitor Cells
HRQoL	Health Related Quality of Life
HSCT	Hematopoietic Stem Cell Transplantation
HST	Highly Specialized Technology
HSV-TK	Herpes Simplex I Virus Thymidine Kinase
HTA	Health Technology Assessment
IAB	Improvement in Actual Benefit *(Amélioration du Service Médical Rendu)*
ICER	Incremental Cost-Effectiveness Ratio
ICER	Institute for Clinical and Economic Review
IM	Intramuscular
IND	Investigational New Drug
INN	International Non-proprietary Names
IP	Intellectual Property
IQWIG	*Institut Für Qualität Und Wirtschaftlichkeit Im Gesundheitswesen*
ITF	Innovation Task Force
ITT	Intention-To-Treat
IV	Intravenous
MA	Marketing Authorisation
MAA	Marketing Authorisation Application
MACI	Matrix-Induced Autologous Chondrocyte Implantation
MEA	Managed Entry Agreement
MPS	Mucopolysaccharidosis
NHL	Non-Hodgkin Lymphoma
NHS	National Health Services

NICE	National Institute for Health and Care Excellence
NIH	National Institutes of Health
OECD	Organisation for Economic Co-Operation and Development
OTAT	Office of Tissue and Advanced Therapies
PDCO	Paediatric Committee
PDUFA	Prescription Drug User Fee Act
PFU	Plaque-Forming Unit
PMBCL	Primary Mediastinal Large B-Cell Lymphoma
PRAC	Pharmacovigilance Risk Assessment Committee
PRIME	Priority Medicines
QALY	Quality Adjusted Life Year
RCT	Randomized Clinical Trial
RMAT	Regenerative Medicine Advanced Therapy
RMP	Risk Management Plan
SGCMPS	General Sub-Directorate of Quality of Medicines and Medical Devices
SMA	Spinal Muscular Atrophy
SMC	Scottish Medicine Consortium
SME	Micro-, Small- and Medium-Sized Enterprise
SPC	Summary of Product Characteristics
TC	Transparency Committee
TDT	Transfusion-Dependent β-Thalassaemia
TEP	Tissue Engineered Product
TLV	Dental and Pharmaceutical Benefits Agency
TPP	Target Product Profile
TTO	Time Trade-Off
UK	United Kingdom
US	United States
VBP	Value Based Pricing
WHO	World Health Organization
ZIN	*Zorginstituut Nederland*

Authors

Eve Hanna holds a doctor of pharmacy degree and a master's degree in pharmaco-epidemiology, and a PhD in health economics. She is an associate researcher at Lyon 1 university, faculty of medicine, unit EA 4129 *parcours de santé systémique (P2S)*. She has successfully completed the European market access university diploma (EMAUD) and is currently a senior lecturer in health technology assessment (HTA) and market access.

Dr. Hanna has been working in the market access consulting field since 2014. She has been involved in advising pharmaceutical companies from majors to start-up companies in market access and HTA strategy for several gene therapies for the United States, Japanese, and European markets. She has been engaged in several multi-HTA early dialogues and parallel scientific advices on gene and cell therapies before phase III or before phase II/III of clinical development. With more than five years of work experience in the field of pricing and market access, she has acquired proficient knowledge in HTA frameworks, pricing, and reimbursement processes and health policy at a global level.

Her area of expertise is gene and cell therapies HTA, funding, and market access; she has several publications in peer-reviewed journals and major conferences all focused on gene and cell therapies. Eve has contributed to three books which mainly focused on gene therapy and orphan drug topics.

Prof. Mondher Toumi is a medical doctor by training. He had earned his MSc in biostatistics and MSc in biological sciences (option pharmacology).Unrelentingly, he also completed his postgraduation in parasitology and in nuclear radiations and health and a PhD in economic sciences. Mondher Toumi has been a professor of health economics and a professor of public health at Aix-Marseille University, and served as a visiting professor at Beijing University.

After working for 12 years as a research manager in the department of pharmacology at the University of Marseille, he joined the Public Health Department in 1993 as an associate. From 1995 to 2008, he worked in the pharmaceutical industry, where he was appointed as the global vice president at Lundbeck A/S in charge of health economics, outcome research, pricing, market access, epidemiology, risk management, governmental affairs, and competitive intelligence.

In 2008, he founded Creativ-Ceutical, an international consulting firm dedicated to support health industries and authorities in strategic decision-making.

In February 2009, he was appointed as a professor at Lyon 1 University in the department of decision sciences and health policies. In the same year, he was appointed as the director of the chair of public health and market access research. Additionally, he created the Market Access Society to promote research and scientific activities around market access, public health, and health economic assessment. Professor Mondher is the editor in chief of the Journal of Market Access and Health Policy (JMAHP).

He is a recognized expert in health economics and an authority on market access and risk management. He has more than 200 scientific publications and more than 500 communications, and has contributed to 15 books, such as a reference book on market access.

chapter one

Introduction to cell and gene therapies concepts and definitions in the US and the EU

The major advances in the field of biotechnology and molecular biology in the twenty-first century have led to a better understanding of pathophysiology of diseases, and to the exploitation of biological components such as nucleic acid sequences, stem cells, for developing new advanced therapies [1]. A new generation of biopharmaceuticals has emerged including a wide and heterogeneous range of cell and gene therapies. These promising therapies aim to prevent or treat chronic or serious life-threatening diseases, previously considered incurable.

Due to these new advanced therapies complexity, the European Parliament and the Council of the European Union (EU) put in place a regulation on November 13, 2007: Regulation (EC) No 1394/2007 [2] on advanced therapy medicinal products and amending Directive 2001/83/EC and Regulation (EC) No 726/2004. This regulation defines and regulates a class of biopharmaceuticals called advanced therapy medicinal products (ATMPs).

In the United States (US), cell and gene therapies were regulated by the Office of Tissue and Advanced Therapies (OTAT), one of the subdivisions of the Centre for Biologics Evaluation and Research (CBER) under the oversight of the Public Health Service Act (PHS Act) and Federal Food, Drug, and Cosmetic Act (FD&C Act) [3]. Regenerative medicine was considered as "human cells, tissues, or cellular or tissue-based products" (HCT/Ps) in Title 21 of the Code of Federal Regulations (CFR) Part 1271.3(d) (21 CFR 1271) [4]. In December 2016, nine years after the EU ATMP regulation, the Food & Drug Administration (FDA) put forth its first interpretation of the "Regenerative Medicine Advanced Therapy" (RMAT) designation program, which was established by section 3033 of the 21st Century Cures Act [5].

1.1 *Introduction to EU regulation (EC) No 1394/2007 of ATMPs*

Since 2001, all medicinal products, biological or chemical origin, have been regulated by the Directive 2001/83/EC [6]. The ATMPs class fall between medical devices and medicinal products. To cover this overlapping area, the EU Parliament implemented a new regulation in November 2007—Regulation (EC) No 1394/2007 [2]—and started its application in December 2008. The aim of this regulation was to "ensure the free movement of these medicines within the European Union (EU), to facilitate their access to the EU market, and to foster the competitiveness of European pharmaceutical companies in the field, while guaranteeing the highest level of health protection for patients" [7]. The ATMP regulation created a single, harmonized framework for ATMPs that ensures universal standards of safety, quality, and efficacy. It completed the legal framework for advanced therapies.

The regulation applies to ATMPs over and above the requirements of the European regulatory framework set out in Directive 2001/83/EC. Standards for the human tissues and cell donation, procurement, testing, processing, storage, and distribution are regulated under Directive 2004/83/EC [8]. Therefore, ATMP regulation was additional and complementary to the other regulations for medicinal products (Figure 1.1).

The two main elements in the new regulation were a centralized marketing authorization (MA) procedure in Europe and a special committee in the European Medicines Agency (EMA) in charge of ATMP scientific assessment and classification—the Committee for Advanced Therapies (CAT). In addition, the regulation covers the supervision and pharmacovigilance of ATMPs in Europe. These regulatory aspects will be detailed in Chapter 3.

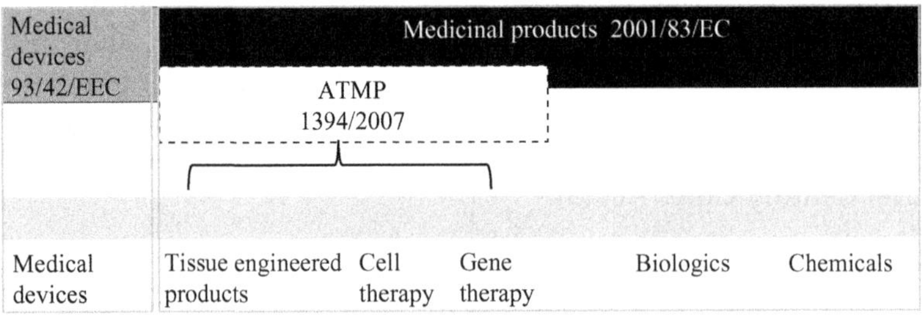

Medical devices 93/42/EEC
Medicinal products 2001/83/EC
ATMP 1394/2007
Medical devices
Tissue engineered products
Cell therapy
Gene therapy
Biologics
Chemicals

Figure 1.1 Advanced therapy medicinal products legislation.

1.1.1 Advanced therapy medicinal products definitions

According to Article 2 of Regulation (EC) No 1394/2007 [2], in addition to the definitions laid down in Article 1 of Directive 2001/83/EC [6] and in Article 3, points (a) to (l) and (o) to (q) of Directive 2004/23/EC [8], ATMP is a class of biopharmaceuticals that encompasses:

- Gene Therapy Medicinal Product (GTMP),
- Somatic Cell Therapy Medicinal Product (sCTMP),
- Tissue-Engineered Product (TEP),
- Combined Therapy Medicinal Product (CATP).

1.1.1.1 Gene therapy medicinal product

As defined in Part IV of Annex I to Directive 2001/83/EC [6], gene therapy medicinal product is a biological product that has the following characteristics:

> contains an active substance which contains or consists of a recombinant nucleic acid used in or administered to human beings with a view to regulating, repairing, replacing, adding or deleting a genetic sequence; its therapeutic, prophylactic or diagnostic effect relates directly to the recombinant nucleic acid sequence it contains, or to the product of genetic expression of this sequence.

An important note to mention is that vaccines against infectious diseases are not considered as GTMP.

There are two approaches for gene therapies: in-vivo and ex-vivo approaches. The in-vivo approach delivers the genetic material directly inside the human body using several vectors such as viral vector, nonviral vector, and naked DNA. The ex-vivo approach consists of genetically modified somatic human cells; the cells are isolated from the human body, the gene is then transferred to the cells, and the cells are reinjected to the body [10].

1.1.1.2 Somatic cell therapy medicinal product and tissue-engineered product

Somatic cell therapy medicinal product means a biological medicinal product that has the following characteristics:

> contains or consists of cells or tissues that have been subject to substantial manipulation so that biological characteristics, physiological functions

> or structural properties relevant for the intended clinical use have been altered, or of cells or tissues that are not intended to be used for the same essential function(s) in the recipient and the donor; presented as having properties for, or is used in or administered to human beings with a view to treating, preventing or diagnosing a disease through the pharmacological, immunological or metabolic action of its cells or tissues.

Somatic cells are all cells in the body except germline cells (sperm and egg).

Tissue-engineered product (TEP) as defined in Article 2 of Regulation (EC) No 1394/20072 [2]:

> contains or consists of engineered cells or tissues, presented as having properties for, or is used in or administered to human beings with a view to regenerating, repairing or replacing a human tissue.

In order to be considered as "engineered" cell or tissue, the product has to fulfill at least one of these two following conditions:

- Substantial manipulation: biological characteristics, physiological functions or structural properties relevant for the intended regeneration, repair or replacement are achieved.
- Not intended to be used for the same essential function in the recipient as in the donor.

TEP may contain cells or tissues of human or animal origin, or both, viable or non-viable. Products containing exclusively non-viable tissues or cells and do not act principally by pharmacological, immunological, or metabolic action, shall not be included in this class. In addition to cells and tissue, TEP may also contain biomolecules, biomaterials, and chemical substances.

The list of manipulation not considered as substantial manipulation as mentioned in Annex I of Regulation (EC) No 1394/2007 [2]:

- Cutting,
- Grinding,
- Shaping,
- Centrifugation,
- Soaking in antibiotic or antimicrobial solutions,
- Sterilization,
- Irradiation,
- Cell separation, concentration or purification,

- Filtering,
- Lyophilization,
- Freezing,
- Cryopreservation,
- Vitrification.

1.1.1.3 Combined therapy medicinal product

Combined advanced therapy medicinal product fulfills the following:

> it must incorporate, as an integral part of the product, one or more medical devices within the meaning of Article 1(2)(a) of Directive 93/42/EEC or one or more active implantable medical devices within the meaning of Article 1(2)(c) of Directive 90/385/EEC, and its cellular or tissue part must contain viable cells or tissues, or its cellular or tissue part containing non-viable cells or tissues must be liable to act upon the human body with action that can be considered as primary to that of the devices referred to.

1.1.1.4 Classification by donor

Advanced therapy medicinal product can be divided into two types: allogeneic and autologous.

- Autologous use is when tissue or cells are derived from the patient himself.
- Allogeneic cells and tissues are derived from a donor whose tissue type closely matches the patient's; the donor can be a family member or not (matched unrelated donor).

An advanced therapy medicinal product containing both autologous and allogeneic cells and tissues is considered as allogeneic use.

1.1.2 Borderline classification

The classifications of ATMPs, drugs, and medical devices may be overlapping in some cases; therefore, the classification may not be immediately obvious.

The regulation specifies that:

- If a product may fall under the categories of sCTMP and TEP, it must be considered as TEP.
- If a product may fall under sCTMP or TEP and GTMP, it must be considered GTMP.

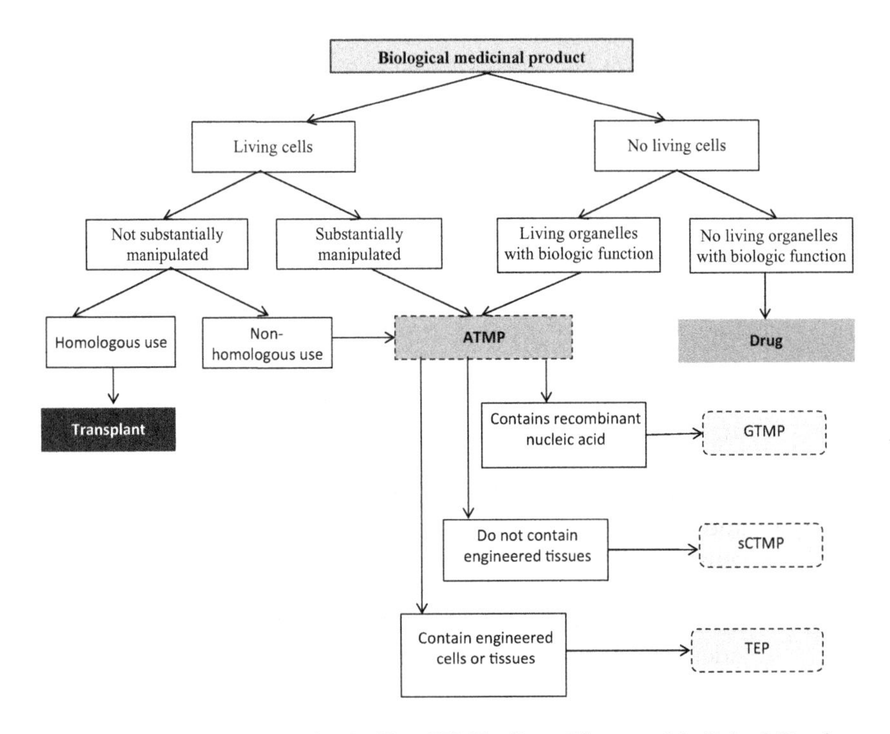

Figure 1.2 Decision tree of ATMPs. GTMP: Gene Therapy Medicinal Product, sCTMP: Somatic Cell Therapy Medicinal Product, TEP: Tissue Engineered Product, CATP: Combined Therapy Medicinal Product. (Data from Pacini, S., *Front. Cell Dev. Biol.*, 2, 2014.)

Furthermore, there can be borderline cases with cosmetics, transplants, or other product types.

Figure 1.2 presents a decision tree that helps to identify if a product should be classified as ATMP or not, and the ATMP class under which it falls.

1.1.3 Hospital exemption

According to Article 28 of Regulation (EC) No 1394/2007 [2], some products are subject to hospital exemptions. Hospital exemption is defined in the regulation as

> ATMPs prepared on a non-routine basis and used within the same Member State in a hospital under the exclusive professional responsibility of a medical practitioner, to comply with an individual medical prescription for a custom-made product for an

individual patient. These products are excluded from the scope of ATMP regulation whilst at the same time ensuring that rules related to quality and safety are not undermined.

Tailor-made products on a nonroutine basis constitute the hospital exemption and do not have to apply for a centralized MA. A member state has to authorize the ATMP manufacture under hospital exemption to ensure the quality of the products [9].

The hospital exemption provides an opportunity for ATMPs to demonstrate clinical proof of concept before initiating the clinical trials. In addition, it may constitute a good opportunity for clinical academic centers developing ATMPs in the early phase of development to benefit from this scheme.

1.1.4 Transitional period

1.1.4.1 Number of ATMPs before the regulation

The European Commission (EC) published in October 2012 a report showing the member states' feedback on the number of ATMPs legally on the market, the number of ATMPs prepared on a routine basis, and the ATMPs that fall under HE and criteria applied for the latter ones by that time [9]. Twenty-seven European countries replied to EC questionnaires, 10 countries had ATMPs legally on the market at that time, and 6 countries had ATMP with HE. Member states have reported 31 ATMPs legally on the EU market prior to the implementation of the ATMP regulation. The same product may have been reported by more than one member state [10].

1.1.4.2 Transitional period timelines

A transitional period has been granted by Article 29 of Regulation (EC) N° 1394/2007 for ATMPs already on the market prior to the regulation in the different member states. During this period, manufacturers had to comply with the ATMP regulation and submit a marketing authorization application for a centralized marketing authorization for their products [11]. The transition period for gene and cell therapies was until December 30, 2011, and for tissue-engineered products was until December 30, 2012.

1.2 ATMPs approved in EU

Eleven years after the regulation implementation, 14 ATMPs have been granted a marketing authorization in the EU, up until December 2019. Approved ATMPs are targeting different diseases and conditions. Five of the ATMPs' marketing authorizations were withdrawn upon their manufacturers' requests: Chondrocelect®, Glybera®, MACI®, Provenge® and Zalmoxis®. (Details on the reasons of withdrawal are discussed in Chapter 4.)

1.2.1 Chondrocelect®

In October 2009, Chondrocelect® was the first approved ATMP in the EU; it is a tissue-engineered product. It is a suspension for implantation that contains characterized viable autologous cartilage cells expanded ex vivo expressing specific marker proteins. It is used in adults with single cartilage defect in the femoral condyle, and it repairs damage to the cartilage in the knee [12].

1.2.2 Glybera®

Glybera®, alipogene tiparvovec, was approved in October 2012 and is a gene therapy medicinal product indicated for the treatment of lipoprotein lipase deficiency [13]. It was designated as an orphan drug EU/3/04/194 on March 8, 2004. It was approved under "exceptional circumstances." The exceptional circumstances mean that the applicant showed that it was not possible to provide comprehensive data on the efficacy and safety of the medicine due to the rarity of the condition, limited scientific knowledge in the area concerned, or ethical considerations. It was under additional monitoring.

1.2.3 MACI®

MACI® is a tissue-engineered product approved in June 2013 for the treatment of cartilage defects. It consists of matrix-applied characterized autologous cultured chondrocytes [14].

1.2.4 Provenge®

Provenge®, sipuleucel-T, is the first somatic cell therapy product approved in September 2013. It contains autologous peripheral blood mononuclear cells activated with PAP-GM-CSF (sipuleucel-T). It is an immunotherapy indicated for the treatment of asymptomatic or minimally symptomatic metastatic castrate-resistant prostate cancer (mCRPC) adults in whom chemotherapy is not yet clinically indicated [15]. It is under additional monitoring which means that it is being monitored even more intensively than other medicines.

1.2.5 *Holoclar®*

Holoclar®, ex vivo expanded autologous human corneal epithelial cells containing stem cells, was approved in February 2015 in the EU. It is a tissue-engineered product, used in the eye to replace damaged cells on the surface (epithelium) of the cornea. It is used in adult patients with moderate to severe limbal stem-cell deficiency caused by burns, including chemical burns, to the eyes [16]. It was granted an orphan designation (EU/3/08/579) on November 7, 2008. It is under additional monitoring. Holoclar® has a conditional approval; this means that the committee's positive opinion was based on data which, while not yet comprehensive, indicate that the medicine's benefits outweigh its risks. Further studies are requested; the approval is renewed on a yearly basis until all obligations are fulfilled, after which the normal approval is issued.

1.2.6 *Imlygic®*

Imlygic®, talimogene laherparepvec, a gene therapy approved in December 2015, is a cancer medicine used to treat adults with unresectable melanoma that is regionally or distantly metastatic (Stage IIIB, IIIC, and IVM1a) with no bone, brain, lung, or other visceral disease [17]. It is under additional monitoring.

1.2.7 *Strimvelis®*

It is a gene therapy approved in 2016, indicated for the treatment of patients with severe combined immunodeficiency due to adenosine deaminase deficiency (ADA-SCID), for whom no suitable human leukocyte antigen (HLA)-matched related stem cell donor is available. It consists of autologous CD34+ enriched cell fraction that contains CD34+ cells transduced with a retroviral vector that encodes for the human ADA cDNA sequence [18]. Strimvelis® was designated as an orphan medicinal product on August 26, 2005. It is under additional monitoring.

1.2.8 *Zalmoxis®*

It is a somatic cell therapy product approved in 2016, indicated as adjunctive treatment in haploidentical hematopoietic stem cell transplantation (HSCT) of adult patients with high-risk hematological malignancies. It consists of allogeneic T cells genetically modified with a retroviral vector encoding for a truncated form of the human low-affinity nerve growth factor receptor (ΔLNGFR) and the herpes simplex I virus thymidine kinase (HSV-TK Mut2) [19]. It was designated an orphan medicinal product on September 17, 2003. Zalmoxis® had a conditional approval and was under additional monitoring.

1.2.9 Spherox®

Spherox® is an autologous medicinal product that contains spheroids of human autologous matrix-associated chondrocytes. It is a TEP, administered by intra-articular implantation for adults to repair the symptomatic articular cartilage defects of the femoral condyle and the patella of the knee (International Cartilage Repair Society [ICRS] grade III or IV) with defect sizes up to 10 cm^2. It was granted a marketing authorization on July 10, 2017 [20].

1.2.10 Alofisel®

Alofisel®, darvadstrocel, is a somatic cell therapy approved in March 2018 for the treatment of complex perianal fistulas in adult patients with nonactive/mildly active luminal Crohn's disease, when fistulas have shown an inadequate response to at least one conventional or biologic therapy. Darvadstrocel contains expanded human allogeneic mesenchymal adult stem cells extracted from adipose tissue (expanded adipose stem cells—eASC). It was designated an orphan drug on October 8, 2009 [21].

1.2.11 Yescarta®

Yescarta®, axicabtagene ciloleucel, is a CD19-directed genetically modified autologous T cell immunotherapy. To prepare Yescarta®, a patient's own T cells are harvested and genetically modified ex vivo by retroviral transduction to express a chimeric antigen receptor (CAR) comprising a murine anti-CD19 single-chain variable fragment (scFv) linked to CD28 and CD3-zeta co-stimulatory domains. The anti-CD19 CAR-positive viable T cells are expanded and infused back into the patient, where they can recognize and eliminate CD19-expressing target cells. Yescarta® is indicated for the treatment of adult patients with relapsed or refractory diffuse large B-cell lymphoma (DLBCL) and primary mediastinal large B-cell lymphoma (PMBCL), after two or more lines of systemic therapy. It was designated an "orphan medicine" (a medicine used in rare diseases) for DLBCL on December 16, 2014, and for PMBCL on October 9, 2015 [22].

1.2.12 *Kymriah®*

Kymriah®, tisagenlecleucel, is an immunocellular therapy containing tisagenlecleucel, autologous T cells genetically modified ex vivo using a lentiviral vector encoding an anti-CD19 CAR. It was approved in the EU on August 23, 2018. Kymriah® is indicated for the treatment of:

- Pediatric and young adult patients up to 25 years of age with B-cell acute lymphoblastic leukemia (ALL) that is refractory, in relapse post-transplant, or in second or later relapse
- Adult patients with relapsed or refractory diffuse large B-cell lymphoma (DLBCL) after two or more lines of systemic therapy

It was designated an "orphan medicine" for B-cell ALL on April 29, 2014, and DLBCL on October 14, 2016 [23].

1.2.13 *Luxturna®*

Luxturna®, voretigene neparvovec, is a gene therapy approved on November 22, 2018. It is indicated for the treatment of adult and pediatric patients with vision loss due to inherited retinal dystrophy caused by confirmed biallelic RPE65 mutations and who have sufficient viable retinal cells.

It was designated an "orphan medicine" for two forms of the disease on various dates (retinitis pigmentosa: July 28, 2015; Leber's congenital amaurosis: April 2, 2012) [24].

1.2.14 *Zynteglo®*

Zynteglo®, autologous CD34+ cells encoding βA-T87Q-globin gene, is a gene therapy approved on May 29, 2019. It is indicated for the treatment of patients 12 years and older with transfusion-dependent β-thalassemia (TDT) who do not have a β0/β0 genotype, for whom hematopoietic stem cell (HSC) transplantation is appropriate but a human leukocyte antigen (HLA)-matched related HSC donor is not available [25].

1.3 *Introduction to regenerative medicine advanced therapy (RMAT) designation*

In December 2016, the 114th US Congress passed the 21st Century Cures Act, a law authorizing $6.3 billion in funding, most of which is allocated for the National Institutes of Health (NIH). The act was passed by both houses of Congress and signed into law by President Obama. The bill encourages late-stage development of advanced therapies, recognizing the potential to address severe unmet medical needs. The 21st Century Cures Act puts the US on the same playing field as European countries that support accelerated of innovative medicines.

This act introduced the concept of regenerative medicine advanced therapy (RMAT). According to Section 3033 of the 21st Century Cures Act [26], a drug is eligible for RMAT designation if:

> The drug is a regenerative medicine therapy, which is defined as a cell therapy, therapeutic tissue engineering product, human cell and tissue product, or any combination product using such therapies or products, except for those regulated solely under Section 361 of the Public Health Service Act and part 1271 of Title 21, Code of Federal Regulations; The drug is intended to treat, modify, reverse, or cure a serious or life-threatening disease or condition; and Preliminary clinical evidence indicates that the drug has the potential to address unmet medical needs for such disease or condition.

On November 16, 2017, the FDA released a draft guidance document that clarified the concept of RMAT established by the 21st Century Cures Act and described the expedited programs available for the development and review of certain regenerative medicine therapies. It highlights the inclusion of "gene therapies, including genetically modified cells, that lead to a durable modification of cells or tissues" in the FDA's interpretation of a "regenerative medicine therapy" that may be eligible for RMAT designation. Therefore, products eligible for an RMAT designation include:

- Cell therapies,
- Gene therapies,
- Therapeutic tissue engineering products,
- Human cell and tissue products,
- Combination products using such therapies or products.

Overall, as of February 2020, the FDA has granted 45 RMAT designations: 11 product designations in 2017 among which 2 were withdrawn, 18 product designations in 2018 among which 2 were withdrawn, and 16 products in 2019 among which 2 were withdrawn (Figure 1.3) [27]. The list of RMAT designated drugs is presented in Table 1.1 [28].

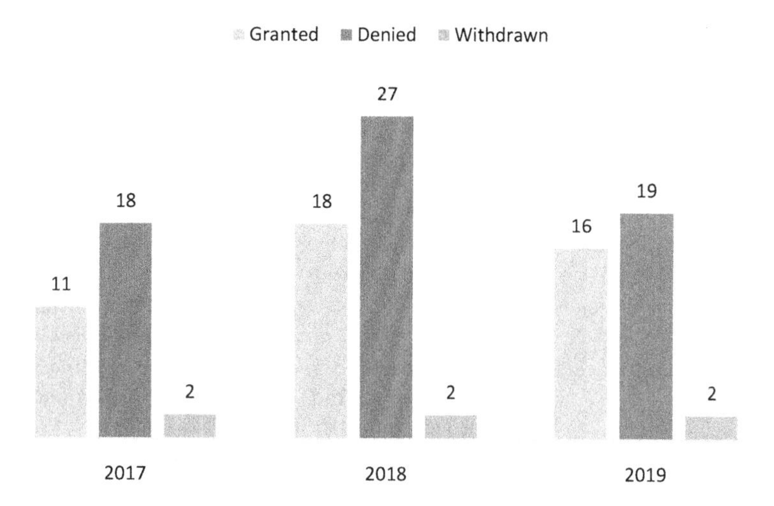

Figure 1.3 Regenerative medicine advanced therapy designation. (Data from FDA website. https://www.fda.gov/vaccines-blood-biologics/cellular-gene-therapy-products/cumulative-cber-regenerative-medicine-advanced-therapy-rmat-designation-requests-received-fiscal [27].)

Table 1.1 List of RMAT designated products

Company	Product	Indication
Asterias Biotherapeutics	AST-OPC1	Spinal Cord Injury (SCI)
Athersys	MultiStem	Ischemic Stroke
Bluebird Bio	LentiGlobin	Severe Sickle Cell Disease
Cellvation Inc. (Fortress Biotech Co.)	CEVA101	Traumatic Brain Injury (TBI)
Humacyte, Inc	Humacyl	Vascular Access for Hemodialysis
Enzyvant	RVT-802	DiGeorge Syndrome
jCyte	jCell	Retinitis Pigmentosa (RP)
Juno Therapeutics	JCAR017	Lymphoma (large B cell NHL)
Kiadis Pharma	ATIR101	Leukemia
Mallinckrodt Pharmaceuticals	Stratagraft®	Thermal Burns
Mesoblast, Ltd.	MPC-150-IM	Heart Failure
Vericel Corporation	Ixmyelocel-T	Dilated Cardiomyopathy
Abeona Therapeutics	EB-101	Recessive Epidermolysis Bullosa

(Continued)

Table 1.1 (Continued) List of RMAT designated products

Company	Product	Indication
Abeona Therapeutics	ABO-102	Sanfilippo Syndrome Type A (MPS IIIA)
Audentes Therapeutics, Inc	AT132	X-linked Myotubular Myopathy
Caladrius Biosciences	CLBS14 (CD34+ cell therapy program)	Refractory Angina
Capricor Therapeutics	CAP-1001	Duchenne Muscular Dystrophy (DMD)
Cellerant Therapeutics, Inc.	Romyelocel-L (human myeloid progenitor cells)	Prevention of Infections During Neutropenia
MiMedx Group	AmnioFix® Injectable	Osteoarthritis (OA) of the Knee
Nightstar Therapeutics	NSR-REP1	Choroideremia (progressive vision loss)
Voyager Therapeutics	VY-AADC	Advanced Parkinson's Disease
AxoGene	Avance	Nerve Injuries
ExCellThera	ECT-001	Blood Cancers
Fortress Biotech	Cellvation's CEVA101	Traumatic Brain Injury
Humacyte		Vascular Access for Hemodialysis
Iovance	Lifileucel	Metastatic Melanoma
MiMedx Group	AmnioFix®	Osteoarthritis
NightStar Therapeutics	NSR-REP1	Choroideremia
Poseida Therapeutics	P-BCMA-101	Multiple Myeloma
Rocket Pharma	RP-L102	Fanconi Anemia
Voyager Therapeutics	VY-AADC	Parkinson's Disease
Mallinckrodt/Stratatech	Stratagraft®	Skin Tissue in Thermal Burns
Talaris Therapeutics	FCR-001	Live donor kidney transplant
AlloVir'	Viralym-M (ALVR105)	Haemorrhagic cystitis (HC) caused by BK virus in adults and children following allogeneic hematopoietic stem cell transplantation (HSCT)
Krystal Biotech	KB103	Dystrophic epidermolysis bullosa

(*Continued*)

Table 1.1 (Continued) List of RMAT designated products

Company	Product	Indication
Fibrocell Science	FCX-007	Recessive dystrophic epidermolysis bullosa
Sangamo Therapeutics	SB-525	Haemophilia A
SanBio	SB-623	Chronic neurological motor deficits secondary to traumatic brain injury
Mustang Bio	MB-107	X-Linked Severe Combined Immunodeficiency
CARsgen	CT-053	Relapsed and/or refractory multiple myeloma
Magenta Therapeutics	MGTA-456	Multiple inherited metabolic disorders

1.4 Approved cell and gene therapies in the US

Until December 2019, 17 cell and gene therapies were approved in the US [29]. Six out of the 16 products are also approved in the EU. This list provides an overview of cell and gene therapies; however, these products may not necessarily be classified as RMAT.

The list of approved cell and gene products, as of December 2019, is:

- Allocord® (HPC Cord Blood)
- Laviv® (Azficel-T)
- MACI® (Autologous Cultured Chondrocytes on a Porcine Collagen Membrane)
- Clevecord® (HPC Cord Blood)
- Gintuit® (Allogeneic Cultured Keratinocytes and Fibroblasts in Bovine Collagen)
- Hemacord® (HPC, cord blood)
- Ducord®, HPC Cord Blood
- HPC, Cord Blood
- HPC, Cord Blood—MD Anderson Cord Blood Bank
- HPC, Cord Blood—LifeSouth
- HPC, Cord Blood—Bloodworks
- Imlygic® (talimogene laherparepvec)
- Kymriah® (tisagenlecleucel)
- Luxturna® (voretigene neparvovec-rzyl)

- Provenge® (sipuleucel-T)
- Yescarta® (axicabtagene ciloleucel)
- Zolgensma® (onasemnogene abeparvovec-xioi).

1.5 Conclusion

An important number of promising advanced therapies are expected to reach the market in the near future. They may be labeled differently between the EU and the US—ATMPs or RMATs or regenerative medicine—but EU and US regulators have acknowledged the specificity of this class and its substantial potential value. Current policy decisions clearly support a disruption in regulators and payers' practices who are tempted to foster access of this new class. Although the future impact of cell and gene therapies is unclear, these therapies are evolving fast from a scientific perspective especially with gene editing becoming reality in the short term.

References

1. Choudhury AR, Kumar N, Sandeep K, et al. Biotechnological potential of stem cells. *Journal of Stem Cell Research & Therapy*. 2017;3(1):212–219. doi:10.15406/jsrt.2017.03.00090.
2. Regulation (EC) No 1394/2007 of the European parliament and of the council of November 13, 2007 on advanced therapy medicinal products and amending directive 2001/83/EC and regulation (EC) no 726/2004. 2007.
3. FDA. OTAT learn. Available from: https://www.fda.gov/biologicsbloodvaccines/newsevents/ucm232821.htm.
4. CFR—Code of Federal Regulations Title 21. Available from: https://www.accessdata.fda.gov/scripts/cdrh/cfdocs/cfcfr/CFRSearch.cfm?CFRPart=1271.
5. Regenerative Medicine Advanced Therapy Designation. Available from: https://www.fda.gov/biologicsbloodvaccines/cellulargenetherapyproducts/ucm537670.htm.
6. The European Parliament and the Council of the European Union, Directive 2001/83/EC. 2001.
7. EMA. ATMP Legal framework. Available from: https://www.ema.europa.eu/en/human-regulatory/overview/advanced-therapies/legal-framework.
8. Directive 2004/23/EC of the European Parliament and of the Council of March 31, 2004 on setting standards of quality and safety for the donation, procurement, testing, processing, preservation, storage and distribution of human tissues and cells. 2004.
9. Van Wilder P. Advanced therapy medicinal products and exemptions to the regulation 1394/2007: How confident can we be? An exploratory analysis. *Frontiers in Pharmacology*. 2012;3:12.

10. European Commission PC. Hospital exemption for ATMPs (implementation of Art 28(2) of Regulation 1394/2007): update on feedback received by the Commission. 2012.
11. Ancans J. Cell therapy medicinal product regulatory framework in Europe and its application for MSC-based therapy development. *Frontiers in Immunology*. 2012;3:253.
12. EMA. Chondrocelect. Available from: https://www.ema.europa.eu/medicines/human/EPAR/chondrocelect.
13. EMA. Glybera. Available from: https://www.ema.europa.eu/en/medicines/human/EPAR/glybera.
14. EMA. MACI. Available from: https://www.ema.europa.eu/en/medicines/human/EPAR/maci.
15. EMA. Provenge. Available from: https://www.ema.europa.eu/en/medicines/human/EPAR/provenge.
16. EMA. Holoclar. Available from: https://www.ema.europa.eu/en/medicines/human/EPAR/holoclar.
17. EMA. Imlygic. Available from: https://www.ema.europa.eu/en/medicines/human/EPAR/imlygic.
18. EMA. Strimvelis. Available from: https://www.ema.europa.eu/en/medicines/human/EPAR/imlygic.
19. EMA. Zalmoxis. Available from: https://www.ema.europa.eu/en/medicines/human/EPAR/zalmoxis.
20. EMA. Spherox. Available from: https://www.ema.europa.eu/en/medicines/human/EPAR/spherox.
21. EMA. Alofisel. Available from: https://www.ema.europa.eu/en/medicines/human/EPAR/alofisel.
22. EMA. Yescarta. Available from: https://www.ema.europa.eu/en/medicines/human/EPAR/yescarta.
23. EMA. Kymriah. Available from: https://www.ema.europa.eu/en/medicines/human/EPAR/kymriah.
24. EMA. Luxturna. Available from: https://www.ema.europa.eu/en/medicines/human/EPAR/luxturna.
25. EMA. Zynteglo. Available from: https://www.ema.europa.eu/en/medicines/.
26. FDA. 21st Century Cures Act. Available from: https://www.fda.gov/regulatoryinformation/lawsenforcedbyfda/significantamendmentstothefdcact/21stcenturycuresact/default.htm.
27. FDA. Cumulative CBER Regenerative Medicine Advanced Therapy (RMAT) Designation Requests Received by Fiscal Year 2019. Available from: https://www.fda.gov/vaccines-blood-biologics/cellular-gene-therapy-products/cumulative-cber-regenerative-medicine-advanced-therapy-rmat-designation-requests-received-fiscal.
28. RMAT List. Available from: https://ipscell.com/rmat-list/.
29. FDA. Approved Cellular and Gene Therapy Products. Available from: https://www.fda.gov/vaccines-blood-biologics/cellular-gene-therapy-products/approved-cellular-and-gene-therapy-products.
30. Pacini S. Deterministic and stochastic approaches in the clinical application of mesenchymal stromal cells (MSCs). *Frontiers in Cell and Developmental Biology*. 2014;2(50).

chapter two

Cell and gene therapies

Genuine products and potential for dramatic value

2.1 Cell and gene therapies specificities

Cell and gene therapies have features that differentiate them from conventional pharmaceuticals; the majority of these therapies are personalized, custom-made, administered once, and provide long-term benefits.

2.1.1 Personalized therapies and manufacturing specificities

Cell and gene therapies are characterized by a high technical complexity and personalized nature that will lead to substantial challenges for their scalable manufacture. The majority of cell and gene therapies are custom-made preparations, involving in some cases the patient's own cells such as autologous cell therapies, autologous chimeric antigen receptor (CAR)-T cell therapies. Therefore, due to its individualized nature, the personalized cell therapy carries a unique set of manufacturing specificities as compared to off-the-shelf-cell therapies or also called allogeneic cell therapies and conventional pharmaceuticals and biologics [1].

There are substantial challenges in the approaches to manufacturing, transportation, and clinical administration of the autologous cell therapies. Patients are part of the manufacturing process of autologous cell therapies which dictates the technical challenges for developing and commercializing safe, effective, and reproducible cell-based therapies. Unlike the production of traditional pharmaceuticals that is based on scale-up production that consists of increasing manufacturing output by increasing the volume processed for each batch, the production of autologous cells requires complex supply logistics and scale-out which consists of replicating the manufacturing line or unit operation to increase the number of batches [2].

Only one batch is used in manufacturing the treatment for a specific patient. This paradigm creates challenges to ensuring the commercial viability of these therapies and good manufacturing practice (GMP) compliance, such as:

- Manufacturing capacity
- Manufacturing reliability
- Manufacturing cost
- Manufacturing flexibility/scalability
- Comparison of test results
- Classification of rooms
- Patient-related logistics
- Regulatory approval of processes and product
- Cleaning
- Cross contamination/mix-ups
- Quality control
- Safety for the operators
- Safety for the patient
- From lab to the market
- Batch release
- Traceability

2.1.2 "Once and done" concept

Unlike conventional therapies, cell and gene therapies typically require a single administration or short-course administration with long-lasting therapeutic benefits. Single administration can have advantages such as avoiding the downsides of long-term treatment including compliance. For example, 11 out of the 14 approved ATMPs in the EU are administered only once (Table 2.1).

2.1.3 Curative or transformative effect with downstream long-term outcomes

Most cell and gene therapies have the potential to change treatment paradigms of genetic diseases resulting from missing or mutated gene(s). Unlike symptomatic treatments, gene and cell therapies aim to correct the underlying cause of a genetic disease, leading to disease eradication or halting the disease progression. Some authors used to refer to cell and gene therapies as "curative therapies," but the definition of "cure" is controversial; it may be defined as one-shot treatment leading to total eradication of disease, and restoration of patients' health. However, in the case of degenerative diseases with irreversible lesions, cell and gene therapies may halt disease progression without restoring the function of tissues irreversibly damaged; in this case, they may not be considered as curative.

Table 2.1 Approved ATMPs frequency of administration

Brand name	Composition	Administration
Chondrocelect®	Autologous chondrocytes	Single dose: 0.8 to 1 million cells/cm^2
Glybera®	Alipogene tiparvovec	Single administration: maximum total dose is 1×10^{12} gc/kg injected in several sites
MACI®	Matrix-applied autologous chondrocytes	Single administration: 0.5 to 1 million cells per cm^2 of defect
Provenge®	Sipuleucel-T	3 doses at approximately 2-week intervals: each dose of 50×10^6 autologous CD54+ cells
Holoclar®	Autologous human corneal epithelial cells	Single treatment of 79,000 to 316,000 cells/cm^2
Imlygic®	Talimogene laherparepvec	Initial dose of 10^6 PFU/mL Subsequent doses of 10^8 PFU/mL every two weeks
Strimvelis®	Autologous CD34+ cells transduced to express ADA	Single dose: 2 and 20 million CD34+ cells/kg
Zalmoxis®	Allogeneic T cells genetically modified	4 doses of $1 \pm 0.2 \times 10^7$ cells/kg with one-month interval
Spherox®	Spheroids of human autologous matrix-associated chondrocytes	Single use: 10–70 spheroids/cm^2 defect
Alofisel®	Darvadstrocel	Single dose consists of 120 million cells to be injected in up to three fistula tracts
Kymriah®	Tisagenlecleucel	Single dose contains 0.2 to 5.0×10^6 CAR-positive viable T cells/kg for patients 50 kg or less, or 0.1 to 2.5×10^8 CAR-positive viable T cells for patients more than 50 kg
Yescarta®	Axicabtagene ciloleucel	Single administration of 2×10^6 CAR-positive viable T cells per kg
Luxturna®	Voretigene neparvovec	Single administration for each eye is 1.5×10^{11} vector genome by subretinal injection
Zynteglo®	Autologous CD34+ cells encoding βA-T87Q-globin gene	Single administration of a minimum of 5.0×106 CD34+ cells/kg

The term "transformative therapies" is nowadays used to refer to these cell and gene therapies that demonstrate a significant potential for substantial and sustained health benefits extending throughout patients' lifetimes [3].

This specificity constitutes a key differentiating factor from conventional therapies administered chronically, the effects of which disappear once the administration stops [4].

2.1.4 *High up-front cost*

Advanced therapies with potential high value are expected to claim high prices. This high price will be charged up-front in the traditional frameworks. Indeed, the price of an approved gene therapy in Europe, Glybera®, was €1.1million per patient, and another gene therapy, Strimvelis®, had a list price that was €594,000 per patient [5] (Table 2.2).

Cell and gene therapies' high prices have grabbed a lot of attention recently; they have become a major topic in policy discussion and

Table 2.2 Available list prices for some cell and gene therapies

Brand name	INN	Recommended dose	Country	Estimated price per patient
Chondrocelect®	Characterized viable autologous cartilage cells	0.8 to 1 million cells/cm^2	Spain	35,000€
			Denmark	43,417€
			Luxembourg	39,674€
			UK	32,026€
			Belgium	39,674€
Glybera®	Alipogene tiparvovec	1×10^{12} gc/kg	Slovakia	1,009.000€ (43,870€ per vial assuming a 70 kg patient)
			Germany	1,022,723€ (43,831.5€ per vial assuming a 70 kg patient)
Provenge®	Sipuleucel-T	50×10^6 autologous CD54+ cells	United States	49,774.62$
			Germany	24,961.5€
Holoclar®	Ex vivo expanded autologous human corneal epithelial cells	79,000–316,000 cells/cm^2	Denmark	86,179€
			Slovakia	88,892€
			Belgium	95,000€
			UK	£80,000

(Continued)

Table 2.2 (Continued) Available list prices for some cell and gene therapies

Brand name	INN	Recommended dose	Country	Estimated price per patient
Imlygic®	Talimogene laherparepvec	Initial: 10^6 PFU/mL Subsequent: 10^8 PFU/mL	UK	35,070€ (assuming 1 year treatment)
			Switzerland	53,506€ (assuming 1 year treatment)
			Germany	28,381€ (assuming 1 year treatment)
Zalmoxis®	Allogeneic T cells genetically modified	$1 \pm 0.2 \times 10^7$ cells/kg	Germany	163,900€
Alofisel®	Darvadstrocel	4 vials of 5 million cells/mL	UK	13,500£ per vial. 54000£ for 4 vials
			France	54,000€
Kymriah®	Tisagenlecleucel	0.6 to 6.0×10^8 cells/kg	United States	475,000€
			France	297,666 €
			UK	282,000£
Yescarta®	Axicabtagene ciloleucel	2×10^6 cells per kg	United States	373,000€
			France	327 000 €

social media. For example, the social group for patients' defense (Ciss, collectif interassociatif de défense des patients) has denounced the excessively high prices claimed by manufacturers for innovative therapies [6].

Many factors drive these high prices claimed by manufacturers: development of high costs and moving findings from bench to bedside, competition with older drugs that have a "virtual monopoly" considered as standard of care, the value delivered by the cell or gene therapy, the disease severity, and high medical unmet needs in the targeted indications [7].

2.2 *Cell and gene therapies value*

2.2.1 *Value for patients*

Advanced therapies hold important promises for patients suffering from chronic diseases with important medical unmet needs [8]. Their potential benefits bring hope to millions of people living with untreatable diseases.

2.2.1.1 *Curing diseases and saving patients from death*

Cell and gene therapies have shown the potential to cure many diseases, some that are partly or fully caused by genetic mutations, such as:

- **Vision disorders such as rare inherited disorder causing vision loss:**
  - Luxturna® (voretigene neparvovec) is the first treatment option for hereditary retinal dystrophy with mutations of the RPE65 gene. It is a severely debilitating disease, with no available treatment characterized by a progressive loss of vision; most patients will become blind.

    Luxturna® is meant for patients with confirmed biallelic mutations of the RPE65 gene and who have sufficient viable retinal cells. It delivers a functional RPE65 gene into the cells of the retina through a single retinal injection, which restores the production pathway for the required enzyme thereby improving the patient's ability to detect light [9].
- **Cardiovascular diseases** [10]:
  - Current treatment advances have not resolved the underlying problem of functional heart tissue loss. Wharton jelly-derived allogeneic mesenchymal stem cells, cultured in vitro, are classified as tissue-engineered products that are in clinical development. They are intended for the treatment of acute myocardial infarction, chronic ischemic heart failure, and no-option critical limb ischemia. These cells may constitute an alternative therapy to ensure self-repair, reversed or attenuated adverse remodeling, and ultimately achieve long-term functional stabilization and improve heart function.
- **Neuromuscular genetic diseases:**
  - Spinal muscular atrophy (SMA) is a genetic disease affecting the part of the nervous system that controls voluntary muscle movement, often leading to early death. AVXS-101 contains the normal version of the SMN gene, which is defective in SMA patients [11].

2.2.1.2 *Survival prolongation*

Advanced therapies have also shown the potential to extend survival in oncological diseases such as CAR-T cells:

- B-cell ALL and DLBCL are aggressive malignancies with significant treatment gaps for patients. The outlook of all patients who relapse from standard of care therapies is poor. DLBCL survival rates are low for the majority of patients due to ineligibility for autologous stem cell transplant (ASCT) or because salvage chemotherapy or ASCT have failed.

  Kymriah®, CAR-T cell approved for B-cell ALL and DLBCL have shown good efficacy; 83% of the patients treated with Kymriah® achieved complete remission (CR) with incomplete blood count

recovery (CRi) within three months of infusion. Results also indicated that 63 patients treated with Kymriah® demonstrated relapse-free survival at six months [12].
- Yescarta® is a second CAR-T cell therapy approved; 72% of patients who received a single infusion of axicabtagene ciloleucel responded to therapy, with 51% achieving a complete response. At one year following infusion, 60% of patients were alive (95% CI: 50.2, 69.2) [13].

2.2.1.3 Halting disease progression

Some advanced therapies may not eradicate the disease but may have the potential to halt several diseases, especially degenerative diseases, such as fatal brain disease, adrenoleukodystrophy, and Parkinson's disease. In other cases, cell and gene therapies have shown the ability to slow down the progression and the harmful effects of some neurodegenerative diseases such as Huntington's disease, Parkinson's disease, kidney disorders, and amyotrophic lateral sclerosis.

- **Example:** Adrenoleukodystrophy is a genetic disease caused by a defective gene on the X chromosome called ABCD1; it triggers a buildup of fatty acids that damage the protective myelin sheaths of the brain's neurons, leading to cognitive and motor impairment.

  A gene therapy in development demonstrated that 15 of 17 patients had stable neurologic functioning more than two years after receiving the gene therapy.

  Newborn screening helps to identify a defective ABCD1 gene before symptoms of adrenoleukodystrophy appear, which creates an opportunity to prevent the disease's degenerative effects before they begin to irreversibly accumulate [14].

2.2.1.4 Improving health-related quality of life (HRQoL)

In addition to improving patients' health, advanced therapies have the potential to improve patients' quality of life as well as caregivers' and families' quality of life [10]. Patients may become independent and may have a positive impact on daily activities, work, and social life. The improvement in HRQoL measures is anticipated both for diseases with no current alternative treatment available, and for diseases with chronic symptomatic treatments. In this latter case, chronic or lifetime treatment burden would be avoided and compliance [primary reason for drug failure] may be improved.

2.2.2 Value for the society

2.2.2.1 Cost saving

Some advanced therapies are expected to replace chronic treatments and, therefore, eliminate costs associated with noncompliance with current treatments. Cell and gene therapies may also decrease the costs of complications and hospitalizations and reduce resource use. They may not only have an impact on direct costs but also on some social, education, and caregiver costs.

2.2.2.2 *Decrease the burden of disease*

Cell and gene therapies have the potential to reduce the clinical, humanistic, and economic burden of chronic diseases. The diseases' prevalence will be significantly reduced with transformative therapies. There is potential for major public impact by slowing or stopping progression of degenerative prevalent conditions or even curing such conditions, e.g., heart failure, Parkinson's disease, rheumatoid arthritis, diabetes, asthma, etc.

2.2.2.3 *Changes in medical paradigm*

Scientists referred to regenerative medicine as the "next revolution in medicine." While conventional therapies used to treat the disease symptoms or consequences of a defect, the regenerative medicine treat the source of the disease; it replaces the defective or missing gene by a normal version of the gene, and restores or replaces cell or tissue.

- Gene therapy can replace a missing or defective gene in a monogenic disease by introducing the gene with viral or nonviral vectors: modifying a patient's DNA to treat disease represents a major shift in medicine. Gene therapy aims to correct the underlying genetic cause of a disease.
    - For example, a gene therapy for hemophilia replaces the faulty gene involved in hemophilia. Clinical trial results showed that nine patients who received this therapy saw substantial increases in the blood-clotting proteins absent in hemophilia.
- Cell replacement therapy is a type of cell therapy; it helps to replace cells that are not functioning, e.g., cell therapy for Parkinson's disease.
    - Parkinson's disease is a chronic and progressive neurodegenerative disorder caused by the gradual loss of dopaminergic neurons in the substantia nigra. Stem cell has the potential to impact the development of disease-modifying treatments for Parkinson's disease by creating dopamine-producing cells from stem cells [15].
- Tissue therapies aim to repair an organ or tissue. Applications of regenerative medicine technology may offer new therapies for injuries, organ failure, or other clinical problems.
    - Regenerative medicine may represent a cure for patients with myocardial infarction, heart failure, and congenital heart diseases [16].

2.2.2.4 Promising prospect

Our study on ATMPs in ongoing development showed that the majority of ATMPs trials were in the early stages (Phase I, I/II: 64.3%, Phase II, II/III: 27.9%, Phase III: 6.9%). Per category of ATMP, we identified 53.6% of trials for somatic cell therapies, 22.8% for tissue-engineered products, 22.4% for gene therapies, and 1.2% for combined products (incorporating a medical device). Disease areas included cancer (24.8%), cardiovascular diseases (19.4%), musculoskeletal (10.5%), immune system and inflammation (11.5%), neurology (9.1%), and others. Of the trials, 47.2% enrolled fewer than 25 patients (Table 2.3) [8].

Similarly, a quarterly report published by the Alliance for Regenerative Medicine [17] identified 977 trials for ATMPs until Q2 2018: 324 in Phase I, 560 in Phase II, and 93 in Phase III:

- 317 (32.44%) trials were for gene therapies
- 314 (32.13%) for gene-modified cell therapies
- 322 (32.95%) cell therapies
- 24 (2.45%) tissue-engineered products

ATMPs represent a fast-growing field of interest. Most of the products are in Phase II of development conditions, which suggests that ATMPs may

Table 2.3 Classification of ATMPs in development

	Phase I and I/II	Phase II and II/III	Phase III	NA	Total
Cancer	146 (24.2%)	69 (26.4%)	18 (27.8%)		**233 (24.8%)**
Cardiovascular diseases	104 (17.2%)	67 (25.7%)	11 (16.7%)		**182 (19.38%)**
Immune system/ inflammation	68 (11.3%)	29 (11.1%)	9 (13.9%)	2 (1.8%)	**108 (11.5%)**
Musculoskeletal system	59 (9.8%)	25 (9.6%)	9 (13.9%)	6 (6.0%)	**99 (10.5%)**
Neurology	61 (10.1%)	23 (8.8%)	1 (1.5%)		**85 (9.1%)**
Gastrointestinal diseases & diabetes	25 (4.1%)	15 (5.7%)	8 (12.3%)	1 (2.1%)	**49 (5.2%)**
Ophthalmology	34 (5.6%)	7 (2.7%)	3 (4.6%)		**44 (4.7%)**
Pulmonology	25 (4.1%)	6 (2.3%)	1 (1.5%)		**32 (3.4%)**
Dermatology	19 (3.1%)	7 (2.7%)	3 (4.6%)		**29 (3.1%)**
Hematology	16 (2.6%)	4 (1.5%)	0		**20 (2.1%)**
Others	47 (7.8%)	9 (3.4%)	2 (3.1%)		**58 (6.2%)**
Total	**604 (64.3%)**	**261 (27.9%)**	**65 (6.9%)**	**9 (0.9%)**	**939 (100%)**

soon reach the market. Targeted therapies have opened the way for new trial methodologies, from which ATMPs could benefit to get early access.

2.2.2.5 *Regenerative medicine could slow the aging process*

Several health conditions are age-related, leading to a decrease in productivity. Europe's population is getting older; the working age population is expected to decrease significantly from 333 million in 2016 to 292 million in 2070 [11]. Regenerative medicines help to cure some age-related conditions such as cartilage repair, leading to slowing the aging process and increasing productivity. Therefore, it may provide an opportunity for a healthier aging.

2.3 *Advanced therapies: a heterogeneous class*

2.3.1 *Value drivers*

ATMP is a heterogeneous class of biopharmaceuticals. This class encompasses different types of therapies targeting various disease areas and providing different values. Cell and gene therapies have the potential to cure or halt several diseases; therefore, their value may be different based on the potential outcomes and several other drivers. The main drivers that define the cell and gene therapies value include:

- Outcomes: cure, prevent, or stop disease progression.
- Availability of alternatives: determine if available therapies are already on the market and used to treat for the targeted disease.
- Disease severity and unmet needs: if the disease is fatal, life-threatening with important unmet clinical needs.
- Disease rarity: the disease may be a rare disease, ultra-rare disease, or a common prevalent disease.
- Frequency of administration: the administration may be once or repetitive.
- Mode of administration: the ATMP may be administered by injections IV, IM, or via complex procedures and surgeries that require hospitalization.
- Population vulnerability: newborns, children, and the elderly are considered more vulnerable populations.
- Availability of a reliable diagnostic test: this is a test with good sensitivity and specificity to detect the disease.

2.3.2 *Different archetypes*

Based on the drivers mentioned above, different cell and gene therapies archetypes with different values may be presented. To illustrate the difference of the value delivered by the different archetypes, some examples are presented in Figure 2.1.

	Archetype 1	Archetype 2	Archetype 3	Archetype 4	Archetype 5
Dosing	Single dose or short-course	Single dose	Short-course treatment	Repetitive treatment	Repetitive treatment
Mode of administration	IV/IM/SC	Procedure	IV/IM	IV/IM	Procedure
Disease Rarity	Rare/or prevalent disease	Rare disease	Prevalent disease	Prevalent disease	Prevalent disease
Disease severity	Fatal or severely disabling disease	Fatal or severely disabling disease	Moderately disabling disease	Moderately disabling disease	Moderately disabling disease
Alternatives	No curative alternatives	Symptomatic treatments	Symptomatic treatments	No available treatments	Available alternatives
Outcome	Curative	Curative	Delay disease progression	Curative	Delay disease progression
Population age (vulnerability)	Children	Adults, elderly	Adults	Adults	Adults
Reliable diagnostic test	Yes	Yes	Yes	No	No

Value

Figure 2.1 Example of ATMPs archetypes

2.4 Conclusion

Unlike conventional therapies, cell and gene therapies—especially autologous cells—are complex, personalized therapies that require special manufacturing processes. Cell and gene therapies are transformative therapies, administered once and delivering lifetime benefits for patients suffering from fatal or chronic diseases. It is a heterogeneous class of different archetypes. For the first time ever, regulators are facing a true pharmaceutical revolution that carries immense hope that some severe fatal conditions may be cured, or controlled, leading to dramatic benefit. Therefore, regulators started to put in place processes and incentives for advanced therapies' developers to continue the investment in this field. Most countries are funding academic research in this field to keep a dominant position. Even though the research field is dominated by the US, several countries are currently engaged to challenge the US domination.

References

1. Advanced therapies investment report 2017. Phacilitate, 2017. Available from: https://ccrm.ca/sites/default/files/pdfs/Investment_for_Advanced_Therapies_Report.pdf.
2. Hourd P CA, Medcalf N, et al. Regulatory challenges for the manufacture and scale-out of autologous cell therapies. In: 2008 CMHSCI, editor. StemBook. 2014.
3. ICER. Value Assessment Methods and Pricing Recommendations for Potential Cures: A Technical Brief Draft Version. 2019.

4. Daniel G, Leschly N, Marrazzo J, McClellan B. Advancing Gene Therapies and Curative Health Care Through Value-Based Payment Reform. 2017. Health Affairs Blog. Available from: https://www.healthaffairs.org/do/10.1377/hblog20171027.83602/full/.
5. Touchot N, Flume M. The payers' perspective on gene therapies. *Nature Biotechnology*. 2015;33(9):902–4.
6. Médicaments innovants: les labos accusés de "marges exorbitantes" 2016. Available from: https://www.sciencesetavenir.fr/sante/medicaments-innovants-les-labos-accuses-de-marges-exorbitantes_30705.
7. Goldstein DA, Stemmer SM, Gordon N. The cost and value of cancer drugs—Are new innovations outpacing our ability to pay? *Israel Journal of Health Policy Research*. 2016;5:40.
8. Hanna E, Remuzat C, Auquier P, Toumi M. Advanced therapy medicinal products: Current and future perspectives. *Journal of Market Access and Health Policy*. 2016;4.
9. Shaberman B. A retinal research nonprofit paves the way for commercializing gene therapies. *Human Gene Therapy*. 2017;28(12):1118–21.
10. Madonna R, Van Laake LW, et al. Position paper of the European society of cardiology working group cellular biology of the heart: Cell-based therapies for myocardial repair and regeneration in ischemic heart disease and heart failure. *European Heart Journal*. 2016;37(23):1789–98.
11. Shorrock HK, Gillingwater TH, Groen EJN. Overview of current drugs and molecules in development for spinal muscular atrophy therapy. *Drugs*. 2018;78(3):293–305.
12. Forsberg MH, Das A, Saha K, Capitini CM. The potential of CAR T therapy for relapsed or refractory pediatric and young adult B-cell ALL. *Therapeutics and Clinical Risk Management*. 2018;14:1573–84.
13. EMA. Yescarta. Available from: https://www.ema.europa.eu/en/medicines/human/EPAR/yescarta.
14. Gene therapy halts progression of cerebral adrenoleukodystrophy in clinical trial. 2017. Available from: https://vector.childrenshospital.org/2017/10/ald-gene-therapy-clinical-trial/.
15. Joana C. Key to Effective Parkinson's Treatment May Lie in Stem Cells, Researchers SayJOANA CARVALHO. 2018. Available from: https://parkinsonsnewstoday.com/2018/07/26/stem-cells-promising-parkinsons-therapy-special-issue/.
16. Alrefai MT, Murali D, Paul A, Ridwan KM, Connell JM, Shum-Tim D. Cardiac tissue engineering and regeneration using cell-based therapy. *Stem Cells Cloning*. 2015;8:81–101. doi:10.2147/SCCAA.S54204.
17. Medicines Afr. Q2 2018 Data Report 2018. Available from: https://alliancerm.org/publication/q2-2018-data-report/.

chapter three

Cell and gene therapies

Regulatory aspects in the US and the EU

3.1 Regenerative medicine regulatory pathway in EU

3.1.1 ATMP specific committee: committee for advanced therapies (CAT)

The Committee for Advanced Therapies (CAT) is a committee within the EMA, established in accordance with Regulation (EC) No 1394/2007 [1].

3.1.1.1 Members

According to Article 21 of Regulation (EC) No 1394/2007 [1], CAT consists of:

- Five members or co-opted members of the Committee for Medicinal Products for Human Use (CHMP) from five member states, with alternates—they are appointed by the CHMP;
- One member and one alternate appointed by each member state not represented among the members and alternates of CHMP;
- Two members and two alternates appointed by the European Commission, in order to represent clinicians;
- Two members and two alternates appointed by the Commission, in order to represent patients' associations.

The committee chair is elected by serving CAT members from among its members. The CAT members and chair are appointed for a three-year renewable period. The list of CAT members and alternates names is published on the EMA website, with their contact details, curriculum vitae, declaration of interests, and confidentiality undertaking.

CAT has the following tasks [1]:

- Providing draft opinion on ATMP quality, safety, and efficacy for final approval by the CHMP;
- Providing advice whether a product falls within the definition of an advanced therapy medicinal product;

- Providing advice on any question related to ATMP at the request of the Executive Director of the Agency or the Commission;
- Assisting scientifically in the elaboration of any documents related to the fulfillment of the objectives of the Regulation;
- Providing scientific expertise and advice for any Community initiative related to the development of innovative medicines and therapies that requires expertise in one of the scientific areas;
- Contributing to the scientific advice procedures.

3.1.1.2 *CAT scientific recommendation on advanced therapy classification*

It is an optional free-of-charge procedure that allows ATMPs manufacturers to seek a scientific recommendation from CAT on whether their product can be classified as an ATMP. In some cases, the manufacturer is uncertain whether the product will fall under medicinal product, medical device, or ATMP. Since the regulatory framework differs significantly between the various healthcare products, it is critical for manufacturers to have clarity about their products' classifications at an early stage of development.

After a 60-day procedure, the CAT issues a nonbinding classification recommendation.

ATMP classification is done by CAT according to Article 17 of Regulation (EC) No 1394/2007. Summaries of scientific recommendations on classification of ATMPs are published on the EMA website [2].

As of April 26, 2019, 313 reports were published: 31 products were not considered as ATMP, 1 product was classified as ATMP but the type of ATMP was not specified, 71 were gene therapies, 73 were somatic cell therapies, and 137 were tissue-engineered product (Figure 3.1).

3.1.2 *Premarketing authorization*

3.1.2.1 *Risk-based approach*

Risk factors are defined as "qualitative or quantitative characteristic that contributes to a specific risk following handling and/or administration of an ATMP" [3]. They include:

- Nature and indication of the ATMP,
- Route of administration and dose,
- Cells origin,
- Phenotype stability,
- Initiation of immune response,

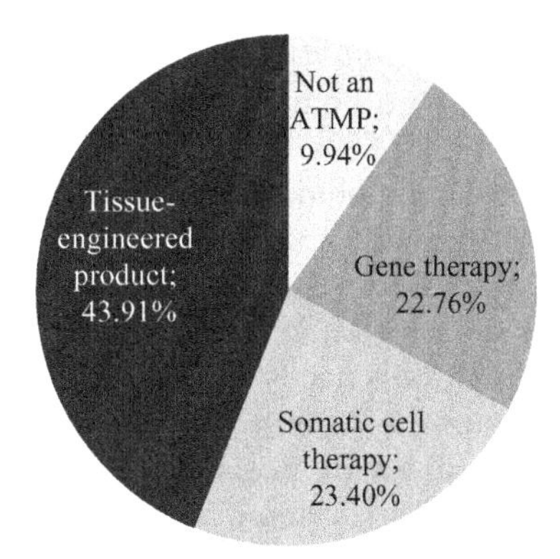

Figure 3.1 Summary of CAT classifications (April 26, 2019).

- Level of cell manipulation,
- Combination with biomolecules or structural biomaterials.

Risk-based approach is used to determine the nature and extent of the quality and (pre)clinical data to be included in the Marketing Authorization Application (MAA). It is a flexible approach that is intended to evaluate and address the risk profile of each ATMP (Directive 2009/120/EC amending Directive 2001/83/EC [4]). Long-term safety issues such as infections, immunogenicity, and device durability for combination ATMPs should also be considered. Furthermore, relevant safety endpoints need to be included in the clinical trials. The available safety, quality, and efficacy data should enable a risk-benefit assessment by CAT.

The risk-based approach was introduced by the EU to create flexibility in the requirements for safety and efficacy of ATMPs [5]. This approach aims to facilitate the science-driven development of ATMPs; it was used by manufacturers to justify deviations from the studies guidelines in 75% of the cases analyzed by Kooijman et al. [5]. The risk analysis can be used to inform the risk management plan (RMP) that needs to be part of an MAA.

The practical implementation of these legal requirements is outlined in the CAT Guideline on the risk-based approach according to Annex I, Part IV of Directive 2001/83/EC applied to advanced therapy medicinal products (Guideline on risk-based approach for ATMPs) [6]. Requirements for pharmacovigilance are laid down in Directive 2001/20/EC, Regulations (EC) 726/2004 and 1394/2007, and the Good Clinical Practice (GCP) guidelines for ATMPs.

3.1.2.2 *Manufacturing quality requirements*

Manufacturing of biologics guidelines apply directly to ATMPs or can be a starting point for the ATMPs development. Good manufacturing practices (GMP) for ATMPs are being developed by EMA as amended in Regulation 1394/2007. Indeed, in December 2016, EMA published the responses of the stakeholders on the ATMP GMP consultation draft.

For biological medicinal product (including ATMPs), the origin and history of starting raw materials need to be described and documented.

The donation, procurement, and testing of human tissues and cells are regulated on an EU-wide basis since 2004 (Directive 2004/23/EC). Procurement supervision systems are set up by member states. Establishments for the donation or testing of cells or tissues need to be accredited, licensed, or authorized.

3.1.2.3 *Nonclinical data*

Given the complexity of ATMPs, a comprehensive nonclinical development program is needed. It is acknowledged that conventional pharmacology and toxicology studies may be inappropriate for these types of products. Nevertheless, a certain number of nonclinical studies are required (Table 3.1).

3.1.2.4 *Clinical data*

ATMPs clinical trials need to adhere to international ethical and quality standards known as Good Clinical Practice (GCP) guidelines. In addition, a detailed GCP special for ATMPs was drawn up by the EC in 2009 to address the special nature of ATMPs.

Studies to be included in clinical dossier—GTMP

Table 3.1 Nonclinical studies required for ATMPs

Nonclinical Studies for Gene Therapies	Nonclinical Studies for Cell Therapy and Tissue Engineering
• Proof of concept	• Proof of concept
• Bio-distribution	• Bio-distribution
• Toxicity assessment	• Toxicity assessment
• Immunogenicity	• Immunogenicity
• Carcinogenicity	• Tumorgenicity
• Vector expression	• Integration of the product
• Insertional mutagenesis germline transmission	• Functional integration
	• Paracrine effects
• Environmental risk assessment GMOs	• Cell differentiation
	• Gonads assessment

- Pharmacokinetic studies: Shedding studies (dissemination of vector/virus through secretions and/or excreta), Dissemination studies—cell tropism, route of administration, target organ/cells, vector type and indication, as well as clinical feasibility and ethical acceptability
- Pharmacodynamics studies
- Dose selection and schedule
- Immunogenicity
- Clinical efficacy
- Clinical safety
- Pharmacovigilance and risk management plan: Gene therapy medicinal products need adequate designed long-term studies to monitor specific efficacy and safety issues [7].

Studies to be included in clinical dossier—sCTMP

For sCTMP, the clinical development program should fulfill the same requirements like other medicinal products. The clinical development plan should include pharmacodynamic studies, pharmacokinetic studies, mechanism of action studies, dose finding studies, and randomized clinical trials in accordance with Directive 2001/20/EC and the existing general guidances and specific guidances for the condition evaluated [8].

3.1.3 Marketing authorization

3.1.3.1 ATMPs guidelines

Several scientific guidelines were developed by EMA to help pharmaceutical industries to prepare the marketing authorization application for ATMPs. Below is a list of ATMPs guidelines available on the EMA website and in the European Pharmacopoeia database [9]:

- The overarching guideline for human gene therapy medicinal products is the note for guidance on the quality, nonclinical, and clinical aspects of gene transfer medicinal products (CHMP/GTWP/671639/2008),
- Guideline on scientific requirements for the environmental risk assessment of gene therapy medicinal products (CHMP/GTWP/125491/06),
- Reflection paper on design modifications of gene therapy medicinal products during development (EMA/CAT/GTWP/44236/2009),
- Reflection paper on quality, nonclinical, and clinical issues relating specifically to recombinant adeno-associated viral vectors (CHMP/GTWP/587488/07),
- ICH Considerations: Oncolytic Viruses (EMEA/CHMP/ICH/607698/2008),

- Guideline on quality, nonclinical, and clinical aspects of medicinal products containing genetically modified cells (CAT/CHMP/GTWP/671639/2008),
- Guideline on the nonclinical studies required before first clinical use of gene therapy medicinal products (EMEA/CHMP/GTWP/125459/2006),
- Guideline on nonclinical testing for inadvertent germline transmission of the gene transfer vectors (EMEA/273974/2005),
- Reflection paper on management of clinical risks deriving from insertional mutagenesis (CAT/190186/2012),
- Guideline on follow-up of patients administered with gene therapy medicinal products (EMEA/CHMP/GTWP/60436/2007),
- The overarching guideline for human cell-based medicinal products is the guideline on human cell-based medicinal products (EMEA/CHMP/410869/2006),
- Reflection paper on stem cell-based medicinal products (EMA/CAT/571134/2009),
- Reflection paper on in vitro cultured chondrocyte containing products for cartilage repair of the knee (EMA/CAT/CPWP/568181/2009),
- Guideline on xenogeneic cell-based medicinal products (EMEA/CHMP/CPWP/83508/2009),
- Guideline on potency testing of cell-based immunotherapy medicinal products for the treatment of cancer (CHMP/BWP/271475/06),
- Reflection paper on clinical aspects related to tissue-engineered products (EMA/CAT/573420/2009).

Latest guidance to clarify regulatory expectations:

- GMP specific to ATMPs: EC guidelines in effect since 2018
- Non-substantially manipulated cell-based ATMPs: Q&A, 2017
- Gene therapy: updated EMA guidelines, 2018
- Safety & efficacy follow-up and risk management: updated EMA guidelines, 2018
- Genetically modified cells: guidelines
- Investigational ATMPs: guidelines
- Comparability for ATMPs: planned for Q4 2019 (CAT work plan)

3.1.3.2 *Marketing authorization application*

The marketing authorization application (MAA) for ATMPs is similar to any medicinal product MAA with technical adaptations. Submission of applications to EMA involves:

- Members of the CHMP
- CAT
- Pharmacovigilance Risk Assessment Committee (PRAC)

The full application should be submitted with a dossier including the following information:

- Pharmaceutical (physicochemical, biological, or microbiological) tests
- Preclinical (toxicological and pharmacological) tests
- Clinical trials
- Any relevant published literature should also be included.

An Electronic Common Technical Document (eCTD) is submitted; it must have the structure presented in Figure 3.2.

Environmental risk assessment: The MAA of an ATMP, like other medicinal products, needs to include an environmental risk assessment (ERA). General guidance for medicinal products for human use is provided (2006). A specific guidance was issued for GTMPs.

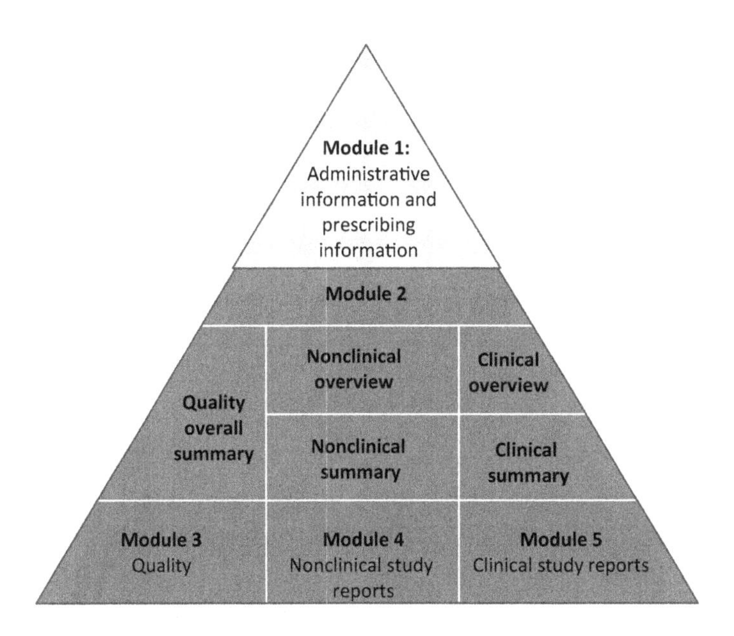

Figure 3.2 Common technical document structure.

3.1.3.3 Centralized marketing authorization procedure

Article 8 of Regulation (EC) No 1394/2007 [1] describes the ATMP evaluation procedure (Figure 3.3):

- CHMP consults the CAT on any scientific assessment needed to have the scientific opinions referred to in Article 5 [2] and [3] of Regulation (EC) No 726/2004. CAT can also be consulted during re-examinations.
- When preparing a draft opinion for final approval by CHMP, the CAT shall endeavor to reach a scientific consensus. If such consensus cannot be reached, the Committee for Advanced Therapies shall adopt the position of the majority of its members.
- The CAT draft opinion shall be sent to the Chairman of the CHMP.
- The CHMP scientific opinion on ATMP may be not in accordance with the draft opinion of the Committee for Advanced Therapies. In this case, CHMP shall annex to its opinion a detailed explanation of the scientific grounds for the differences.

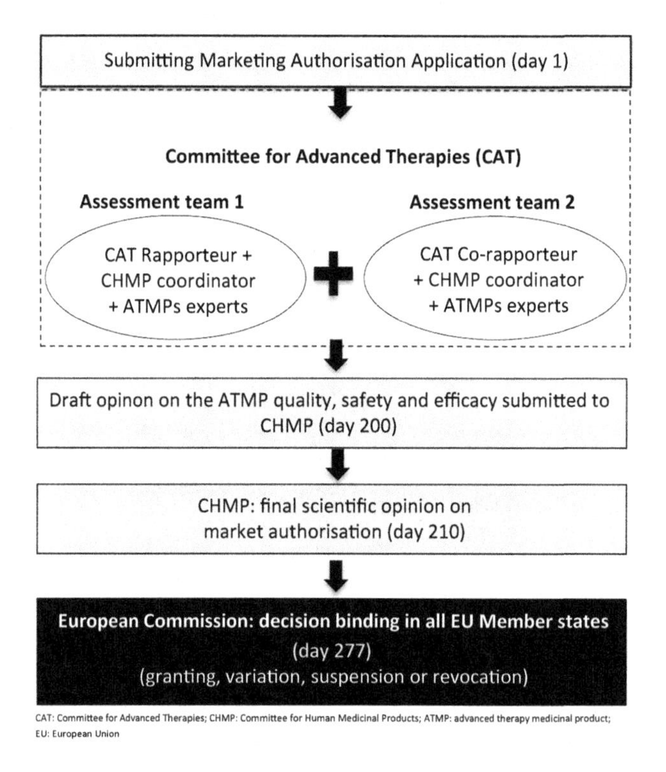

Figure 3.3 Marketing authorization procedure.

- The Pharmacovigilance Risk Assessment Committee ("PRAC") provides recommendations to the CHMP on pharmacovigilance matters.
- The Paediatric Committee ("PDCO") intervenes on aspects related with the obligations imposed under Regulation (EC) No 1901/2006 of the European Parliament and of the Council.
- The Committee on Orphan Medicinal Products ("COMP") provides scientific opinions to the Commission on aspects related to the application of the orphan incentives (this committee is only involved therefore if the applicant seeks orphan status).

3.1.3.4 Marketing authorization timelines

The procedure starts after the application validation by EMA and when all CAT (Co)-Rapporteurs and CHMP Coordinators have received the dossier and all additional information requested during validation.

Table 3.2 shows the standard timetable for the evaluation of an ATMP under the centralized application [10].

After adoption of a CHMP opinion, the applicant is requested to provide the following annexes to EMA:

- Summary of product characteristics
- Final translations of summary of product characteristics, labeling, and package leaflet in the 20 languages

The European Commission makes the final decision.

3.1.3.5 Marketing authorization fees

The fee for marketing authorization shall be reduced by 50% if the applicant is a hospital or a micro-, small-, and medium-sized enterprise (SME) and able to prove the interest of the ATMP concerned. This also accounts for fees charged by the agency for post-authorization activities in the first year following the granting of the marketing authorization for the ATMP [1]. The general marketing authorization fees paid for EMA are summarized in Table 3.3 [11].

The strength is defined as follows [12]:

> For single-dose preparations, total use, the strength is defined as the amount of active substance per unit dose
>
> For multi-dose preparations, the strength is defined as the concentration expressed as the amount of active substance per ml, per puff, per drop, per kg, per m^2, as appropriate.

Table 3.2 Standard timetable for the evaluation of an advanced therapy medicinal product

Day	Action	Responsibilities
1	Start of the procedure	
80	CAT (co)-rapporteurs assessment report(s) are sent to CHMP Coordinators, CAT, and CHMP members and EMA. EMA responds to the applicant with preliminary conclusions that do not represent the position of CAT/CHMP.	CAT (co)-rapporteurs
100	Comments from CHMP Coordinators, members of CAT, and the CHMP (including peer reviews)	CHMP Coordinators, members of CAT, and the CHMP
115	Draft list of questions from CAT (co)-rapporteur, as discussed with the CHMP Coordinators, peer reviewers, CAT, and CHMP members and EMA	CAT (co)-rapporteurs
120	CAT adoption of the list of questions, overall conclusions, and review of the scientific data to be sent to the applicant by EMA	CAT
120	GMP/GLP/GCP Inspection procedure starts [Clock stop]	CAT
121	Submission of the responses including revised SPC, labeling and package leaflet texts in English [restart of the clock]	Applicant
150	Joint response assessment report from CAT (co)-rapporteurs received by CHMP coordinators, CAT and CHMP members and the EMA EMA respond to the applicant with preliminary conclusions that do not represent the position of CAT/CHMP	CAT (co)-rapporteurs
160	Deadline for comments from CAT and CHMP members to be sent to CAT (co)-rapporteurs, CHMP coordinators, EMA, and other CAT and CHMP members	CAT and CHMP members
170	CAT discussion and decision on the need for adoption of a list of outstanding issues and/or oral explanation by the applicant (*If oral explanation is needed, the clock is stopped to allow the applicant to prepare the oral explanation.)	CAT

(*Continued*)

Table 3.2 (Continued) Standard timetable for the evaluation of an advanced therapy medicinal product

Day	Action	Responsibilities
171	CAT oral explanation Discussion on the draft opinion and identification of the recommendations for MA/refusal, which will be transmitted to CHMP Final draft of English SPC, labeling/package leaflet + where needed an updated RMA plan and traceability system	CAT
180	CHMP discussion on the Grounds for approval/ refusal as adopted by CAT	CHMP
200	Draft opinion and draft Assessment Report transmission to the CHMP	CAT
210	Adoption of CHMP Opinion and CHMP Assessment Report	CHMP

CAT: Committee of Advanced Therapies, CHMP: Committee for Medicinal Products for Human Use

Table 3.3 Payable fees

Full Marketing Authorization		Fees
Basic	Single strength associated with one pharmaceutical form and one presentation (basic fee)	278,800€
Additional	For each additional strength or pharmaceutical form including one presentation, submitted at the same time as the initial application for authorization (additional fee)	+28,000€
Additional	For each additional presentation of the same strength and pharmaceutical form, submitted at the same time as the initial application for authorization (additional fee)	+7,000€

3.1.4 Regulatory early access tools and development support

Regulation (EC) No 726/2004 [13] includes a number of provisions and early access tools aimed at fostering patients' access to new medicines that fulfill important unmet clinical needs [14]. In view of the promising clinical potential of ATMPs, ATMPs can also benefit from those tools.

3.1.4.1 Accelerated assessment

This procedure is applicable for therapies of major interest in particular innovation; it reduces the marketing authorization application assessment to 150 days instead of 210 days. The request with the claim that the medicinal product addresses to a significant extent the unmet medical needs must be submitted 2–3 months before MAA submission.

3.1.4.2 *Conditional MA*

This procedure applies to medicinal products for seriously debilitating diseases or life-threatening diseases, emergency situations, and orphan drugs. The product has to prove a positive risk-benefit balance, fulfillment of an unmet medical need. It consists of an earlier authorization on the basis of incomplete clinical data. Manufacturers provide complementary data in a time frame after authorization. It is requested at submission of MAA.

3.1.4.3 *CHMP Compassionate use opinion*

Unauthorized medicinal products indicated for chronically debilitating or life-threatening diseases for a group of patients with no alternative treatments. A member state (MS) can decide to make the medicinal product available for compassionate use, available to patients through national patients' access programs (prior to a marketing authorization). The competent authority of an MS must notify the EMA.

3.1.4.4 *Development support: PRIME*

The PRIME was launched by EMA to ensure rapid market access to promising medicines where a major public health interest presents significant unmet medical need. It is applicable to therapies of major interest in particular innovation. PRIME is a voluntary scheme that aims to optimize the generation of robust clinical data and accelerate authorization application assessments. It helps with the following:

- Accelerated assessment potential identification,
- Early rapporteur appointment,
- Reinforced scientific and regulatory support from the SAWP/CHMP, other relevant scientific committees and EMA,
- Dedicated contact person within EMA.

The manufacturer applies to this procedure during development after having preliminary clinical evidence and identification of the unmet medical need and its magnitude. Until March 2019, eight ATMPs have been granted PRIME designation, evidencing its value (Table 3.4).

Until April 2019, two ATMPs were approved with the PRIME scheme; however, there are an important number of ATMPs in ongoing development that benefit from PRIME (Figure 3.4). Forty-three percent of designated PRIME products are ATMPs (March 2019).

Table 3.4 ATMPs granted PRIME designation (April 2019)

Name[a]	Therapeutic Indication	Type of Data Supporting Request	Type of Applicant	Date of Granting PRIME Eligibility
Adeno-associated viral vector containing factor IX gene variant (PF-06838435/SPK-9001)	Treatment of hemophilia B	Nonclinical + Clinical exploratory	Other	February 23, 2017
Adeno-associated viral vector serotype 5 containing a B-domain deleted variant of human coagulation factor VIII gene (BMN 270)	Treatment of hemophilia A	Nonclinical + Clinical exploratory	Other	January 26, 2017
Adeno-associated viral vector serotype 5 containing human factor IX gene or variant (AMT-060, AMT-061)	Treatment of severe hemophilia B	Nonclinical + Clinical exploratory	SME	April 21, 2017
Adeno-associated viral vector serotype 8 containing the human MTM1 gene (AT132)	Treatment of X-linked myotubular myopathy	Nonclinical + Clinical exploratory	SME	May 31, 2018
Adenovirus associated viral vector serotype 8 containing the human CNGB3 gene (AAV2/8-hCARp.hCNGB3)	*Treatment of achromatopsia associated with defects in CNGB3*	*Nonclinical + Tolerability first in man*	*SME*	February 22, 2018
Autologous CD34+ cells transduced with lentiviral vector encoding the human beta globin gene (OTL-300)	Treatment of transfusion-dependent β-thalassemia	Nonclinical + Clinical exploratory	SME	September 20, 2018

(*Continued*)

Table 3.4 (Continued) ATMPs granted PRIME designation (April 2019)

Name[a]	Therapeutic Indication	Type of Data Supporting Request	Type of Applicant	Date of Granting PRIME Eligibility
Autologous CD4 and CD8 T cells transduced with lentiviral vector containing an affinity-enhanced T cell receptor to target the cancer-testis tumor antigen NY-ESO-1 (NY-ESO-1c259T)	Treatment of HLA-A*0201, HLA-A*0205, or HLA-A*0206 allele positive patients with inoperable or metastatic synovial sarcoma who have received prior chemotherapy and whose tumor expresses the NY-ESO-1 tumor antigen	Nonclinical + Clinical exploratory	Other	July 21, 2016
Autologous CD4+ and CD8+ T cells Expressing a CD19-Specific Chimeric Antigen Receptor (JCAR017)	Treatment of relapsed/refractory diffuse large B-cell lymphoma	Nonclinical + Clinical exploratory	Other	December 15, 2016
Autologous hematopoietic stem cells transduced with lentiviral vector Lenti-D encoding the human ATP-binding cassette, subfamily D (ALD), member 1 (ABCD1) from cDNA	Treatment of cerebral adrenoleukodystrophy	Nonclinical + Clinical exploratory	Other	July 26, 2018
Autologous human T cells genetically modified ex vivo with a lentiviral vector encoding a chimeric antigen receptor (CAR) for B-cell maturation antigen (BCMA) (JNJ-68284528)	Treatment of adult patients with relapsed or refractory multiple myeloma, whose prior regimens included a proteasome inhibitor, an immunomodulatory agent and an anti-CD38 antibody and who had disease progression on the last regimen	Nonclinical + Clinical exploratory	Other	March 3, 2019

(Continued)

Table 3.4 (Continued) ATMPs granted PRIME designation (April 2019)

Name[a]	Therapeutic Indication	Type of Data Supporting Request	Type of Applicant	Date of Granting PRIME Eligibility
Autologous T cells transduced with retroviral vector encoding an anti-CD19 CD28/CD3-zeta chimeric antigen receptor (KTE-X19)	Treatment of adult patients with relapsed or refractory mantle cell lymphoma	Nonclinical + Clinical exploratory	Other	May 31, 2018
Autologous T lymphocyte-enriched population of cells transduced with a lentiviral vector encoding a chimeric antigen receptor targeting human B cell maturation antigen with 4-1BB and CD3-zeta intracellular signaling domains (bb2121)	Treatment of relapsed and refractory multiple myeloma patients whose prior therapy included a proteasome inhibitor, an immunomodulatory agent and an anti-CD38 antibody	Nonclinical + Clinical exploratory	Other	November 9, 2017
Genetically modified replication-incompetent herpes simplex virus-1 expressing collagen VII (KB103)	Treatment of dystrophic epidermolysis bullosa	Nonclinical + Clinical exploratory	SME	March 28, 2019
Lonafarnib	Treatment of hepatitis D virus infection	Nonclinical + Clinical exploratory	SME	December 13, 2018
Lumasiran	Treatment of primary hyperoxaluria Type 1	Nonclinical + Clinical exploratory	Other	March 22, 2018
MV-CHIK vaccine	Prevention of Chikungunya fever	Nonclinical + Clinical exploratory	SME	May 31, 2018
Mycobacterium tuberculosis (MTBVAC)	*Active immunization against tuberculosis disease in newborns, adolescents and adults*	*Nonclinical + Tolerability first in man*	*SME*	June 28, 2018

(*Continued*)

Table 3.4 (Continued) ATMPs granted PRIME designation (April 2019)

Name[a]	Therapeutic Indication	Type of Data Supporting Request	Type of Applicant	Date of Granting PRIME Eligibility
Nangibotide (LR12)	*Treatment of septic shock*	*Nonclinical + Tolerability first in man*	*SME*	November 9, 2017
Olipudase alfa	Treatment of non-neurological manifestations of acid sphingomyelinase deficiency	Nonclinical + Clinical exploratory	Other	May 18, 2017
Purified inactivated Zika virus vaccine (TAK-426)	Active immunization for the prevention of disease caused by Zika virus	Nonclinical + Clinical exploratory	Other	March 28, 2019
Rapastinel	Adjunctive treatment of major depressive disorder	Nonclinical + Clinical exploratory	Other	May 18, 2017
Recombinant adeno-associated viral vector serotype S3 containing codon-optimized expression cassette encoding human coagulation factor IX variant (FLT180a)	Treatment of hemophilia B	Nonclinical + Clinical exploratory	SME	February 28, 2019
Recombinant vesicular stomatitis virus with envelope glycoprotein replaced by Zaire ebolavirus (Kikwit strain) glycoprotein	Vaccination against Ebola (Zaire strain)	Nonclinical + Clinical exploratory	Other	June 23, 2016
Risdiplam	Treatment of 5q spinal muscular atrophy	Nonclinical + Clinical exploratory	Other	December 13, 2018
Seladelpar (MBX-8025)	Treatment of primary biliary cholangitis	Nonclinical + Clinical exploratory	SME	October 13, 2016

(Continued)

Table 3.4 (Continued) ATMPs granted PRIME designation (April 2019)

Name[a]	Therapeutic Indication	Type of Data Supporting Request	Type of Applicant	Date of Granting PRIME Eligibility
Setmelanotide	Treatment of obesity and the control of hunger associated with deficiency disorders of the MC4R receptor pathway	Nonclinical + Clinical exploratory	SME	June 28, 2018
Setrusumab (Recombinant humanized monoclonal IgG2 lambda antibody against human sclerostin, BPS804)	Treatment of osteogenesis imperfecta types I, III, and IV	Nonclinical + Clinical exploratory	SME	November 9, 2017
Synthetic 47-amino-acid N-myristoylated lipopeptide, derived from the preS region of hepatitis B virus	Treatment of chronic hepatitis D infection	Nonclinical + Clinical exploratory	SME	May 18, 2017
Tabelecleucel (Allogeneic Epstein-Barr virus-specific cytotoxic T lymphocytes, ATA129)	Treatment of patients with Epstein-Barr virus-associated post transplant lymphoproliferative disorder in the allogeneic hematopoietic cell transplant setting who have failed on rituximab	Nonclinical + Clinical exploratory	Other	October 13, 2016
Tasadenoturev (Adenovirus serotype 5 containing partial E1A deletion and an integrin-binding domain, DNX-2401)	Treatment of recurrent glioblastoma in patients for which a gross total resection is not possible or advisable, or for those who refuse further surgery	Nonclinical + Clinical exploratory	Other	July 21, 2016
Vocimagene amiretrorepvec	Treatment of high-grade glioma	Nonclinical + Clinical exploratory	SME	July 20, 2017
Voxelotor (GBT440)	Treatment of sickle cell disease	Nonclinical + Clinical exploratory	Other	June 22, 2017

[a] Name of the active substance, INN, common name, chemical name or company code.

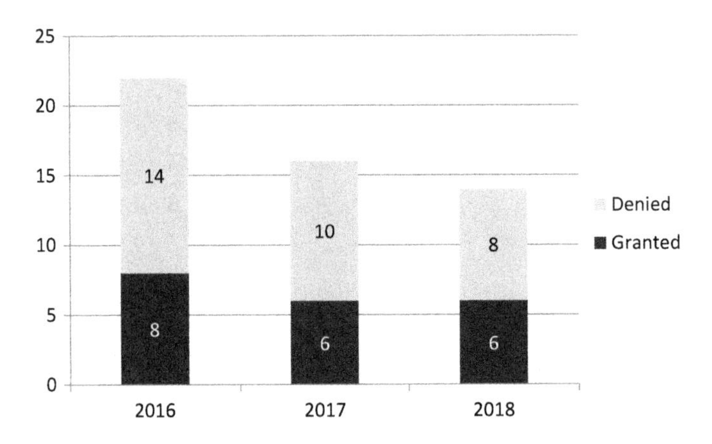

Figure 3.4 PRIME eligibility recommendations for ATMPs (March 2019).

3.1.4.5 *Innovation task force (ITF)*

The ITF is an early dialogue between applicants and a multidisciplinary group (scientific, regulatory, and legal competences) to proactively identify issues related to emerging innovative therapies and technologies. It helps to prepare for regulatory processes. Applicants are mainly SME and academics. This dialogue is free of charge.

3.1.4.6 *SME office*

The SME are defined as follows (Commission Recommendation 2003/361/EC) [15]:

- Micro enterprise: Annual work unit < 10, Annual turnover ≤ €2M Or Annual balance sheet total ≤ €2M.
- Small enterprise: Annual work unit < 50, Annual turnover ≤ €10M Or Annual balance sheet total ≤ €10M.
- Medium enterprise: Annual work unit < 250, Annual turnover ≤ €50M Or Annual balance sheet total ≤ €43M.

EMA offers incentives for SME. Within the EMA, dedicated personnel are present to provide administrative and procedural assistance, monitor MAA, and organize training sessions for SME.

3.1.4.7 *Scientific advice and protocol assistance*

Scientific advice: EMA provides advice on the development and trials of a medicine in response to company questions. A 90% reduction for

Table 3.5 Summary of financial incentives for ATMP developers and/or SME

	Fee Reduction	
Type of Fee	ATMP Developer	SME
Certification procedure for SME	↓90%	N/A
Any scientific advice	↓65%	↓90% (↓100% for designed orphan drug)
Inspection (pre-authorization)	—	↓90%, Deferral of total applicable fee[a]
Application for marketing authorization	↓50% (if applicant is a hospital)	Deferral of total applicable fee[a] Conditional fee exemption, where EMA scientific advice is followed and MA application is not successful

[a] Up to 45 days after the date of the notification of the final decision on the marketing authorization or of the withdrawal of the application.

SMEs and a 65% reduction for other applicants shall apply for such scientific advice or for advice on the conduct of the tests and trials necessary to demonstrate the quality, safety, and efficacy of the ATMPs (Table 3.5).

The manufacturer may request advice on the design and conduct of the pharmacovigilance and of the risk management system.

Parallel scientific advice with health technology assessment (HTA) bodies: This helps to gain feedback from regulators and HTA bodies simultaneously, on the evidence that both parties request to determine a medicine's benefit-risk balance and value.

3.1.5 Post-authorization requirements

Article 14 of Regulation (EC) No 1394/2007 required the MA holder to submit in the MA application "the measures envisaged ensuring the follow-up of efficacy of advanced therapy medicinal products and of adverse reactions thereto." In this same article, EMA was requested to prepare guidelines for the post-authorization follow-up of efficacy and adverse reactions, and risk management. EMA has issued a detailed guideline on this matter (Guideline on Safety and Efficacy Follow-up—Risk Management of ATMPs 2008) [16]. ATMPs need to comply with post-marketing requirements such as other medicinal products; safety and effectiveness data are continuously collected, evaluated, and reported to allow a benefit-risk management.

The risk management is "a set of pharmacovigilance activities and interventions designed to identify, characterize, prevent or minimize risks relating to medicinal products, including the assessment of the effectiveness of those interventions" (EMEA/CHMP/96268/2005). Risk management is a cycle that starts by risk identification from clinical trials, literature review, and epidemiological studies, followed by risk characterization and assessment, then risk minimization and effectiveness measurement [17]. The safety and efficacy follow-up systems should be part of the risk management plan.

There are specific rules for ATMPs with unique safety and efficacy concerns, as well as consideration of endpoints that MA holders need to comply with when designing post-authorization follow-up studies. The guideline on safety and efficacy follow-up-risk management of advanced therapy medicinal products [16] detailed all additional requirements for the Risk Management Plan (RMP) for market authorization holders of approved ATMPs: Safety specifications, Pharmacovigilance plan, Evaluation of the need for efficacy follow-up, Risk Minimization plan, and Efficacy follow-up plan.

3.1.6 *Incentives of the ATMPs regulation*

A number of incentives to manufacturers have been introduced by Regulation (EC) No 1394/2007, especially for SME. These incentives are:

- Scientific advice on the design;
- Scientific advice fees reduction of 90% for SMEs and 65% for other applicants;
- Scientific recommendation on advanced therapy classification;
- Certification of quality and nonclinical data: special to SMEs, these organizations can submit the available quality and nonclinical data for their ATMP to CAT for scientific evaluation and certification. This evaluation applies the same scientific standards and technical requirements as during the assessment of MAA, just at an earlier stage of development. This helps SMEs to attract capital and to facilitate the ATMP's development through the clinical stage and marketing authorization.
- Reduction of the fees for marketing authorization by 50% for hospitals and SMEs if a public health interest in the ATMP can be proven; furthermore, a 50% fee reduction is also granted for post-authorization activities in the first year following the MA.

Manufacturers can also benefit from other incentives such as scientific advice or protocol assistance that are not special for ATMPs.

3.2 Regenerative medicine regulatory pathway in the US

In November 2017, CBER published draft guidance with nonbinding recommendations for industries developing regenerative medicines. The programs described in this guidance are intended to facilitate development and review of regenerative medicine therapies intended to address unmet medical need in those with serious conditions.

Sponsors of regenerative medicine therapies, including RMATs designated therapies, have the possibility to pursue either accelerated approval or traditional approval. The selection of the pathway will depend on the clinical trials design and results providing the primary evidence of effectiveness. CBER encourages sponsors to consult with the agency early in development if interested in pursuing accelerated approval. These interactions can be used to discuss if accelerated approval is appropriate, proposed surrogate or intermediate clinical endpoints, clinical data development plan, and clinical trial design.

3.2.1 FDA expedited programs

Regenerative medicines intended to treat, modify, reverse, or cure serious conditions can benefit from FDA expedited programs, including fast-track designation, breakthrough therapy designation, RMAT designation, accelerated approval, and priority review designation, if they meet the criteria for such programs.

RMAT designation is not the same as an approval and does not change the statutory standards for demonstration of safety and effectiveness needed for marketing approval.

3.2.1.1 Fast-track designation

An investigational new drug (IND) can benefit from a fast-track designation if:

- it is intended to treat a serious condition
- it has the potential to address an unmet medical need
- for which nonclinical or clinical data demonstrate the potential to address an unmet medical need.

The advantages of fast track designation include:

- Actions to facilitate the development of the product
- Expedited review of the product. The product could be eligible for priority review if supported by clinical data at the time of marketing application submission.

3.2.1.2 Breakthrough therapy designation

IND qualifies for a breakthrough therapy designation if:

- it is intended to treat a serious condition, and
- for which preliminary clinical evidence demonstrates substantial improvement over available therapies on one or more clinically significant endpoints.

The level of evidence required for this designation is higher than fast-track designation. The decision cannot be based on nonclinical data.

Advantages of this designation include the benefits of fast-track designation in addition to intensive FDA guidance on efficient drug development, organizational commitment in facilitating the product development program.

3.2.1.3 Regenerative medicine advanced therapy designation

An IND is eligible for RMAT designation if:

- It meets the definition of regenerative medicine therapy (see Chapter 1);
- It is intended to treat, modify, reverse, or cure a serious condition; and
- Preliminary clinical evidence indicates that the regenerative medicine therapy has the potential to address unmet medical needs for such condition.

The preliminary clinical evidence required would be obtained from clinical investigations specifically conducted to assess the effects of the therapy on a serious condition. At the initial stages of product development, randomized clinical trials versus active control may not be possible. In some cases, clinical evidence obtained from studies with appropriately chosen historical controls may be sufficient.

To assess whether the preliminary clinical evidence is sufficient to support RMAT designation, the factors used are:

- the rigor of data collection;
- the nature and meaningfulness of the outcomes;
- the number of patients or subjects;
- the number of sites contributing to the data;
- the severity, rarity, or prevalence of the condition.
- bias in the study design, treatment assignment, or outcome assessment.

RMAT designation does not require evidence to indicate that the drug may offer a substantial improvement over available therapies.

Advantages of the RMAT designation include all of the benefits of the fast-track and breakthrough designation programs, including early interactions with sponsors. According to Section 506(g) [5] of the FD&C Act the early interactions may be used to discuss potential surrogate or intermediate endpoints to support accelerated approval.

Product sponsor benefits:

- Interactions with the FDA.
- Priority review and accelerated approval.
- Flexibility in clinical data requirements: the number of clinical sites used and the possibility to use patient registry data and other sources of "real-world" evidence for post-approval studies.

3.2.1.4 *Priority review designation*

A product that receives fast-track, breakthrough, or RMAT designation may be eligible for priority review, if supported by clinical data at the time the marketing application is submitted.

A decision about granting priority review is made within 60 calendar days of receipt of the marketing application. If priority review is granted, CBER has a six-month goal for reviewing the biologics license application (BLA).

3.2.1.5 *Accelerated approval*

FDA grants accelerated approval to drugs that:

- treat serious conditions,
- provide a meaningful therapeutic advantage over available therapies,
- demonstrate an effect on: (1) a surrogate endpoint that is reasonably likely to predict clinical benefit, or (2) a clinical endpoint that can be measured earlier than irreversible morbidity and mortality that is reasonably likely to predict an effect on irreversible morbidity and mortality or other clinical benefit (i.e., an intermediate clinical endpoint).

RMATs may be eligible for accelerated approval based on:

- previously agreed-upon surrogate or intermediate endpoints that are reasonably likely to predict long-term clinical benefit, or
- reliance upon data obtained from a meaningful number of sites.

3.2.2 Application for RMAT designation

A sponsor should submit a request to CBER either with a new IND application or in an IND amendment. The request should contain a summary of information that supports the designation, including:

- Rationale for the IND meeting the definition of a regenerative medicine therapy;
- Discussion to support that the disease or condition, or the aspect of the disease or condition, that the product is intended to treat is serious;
- Summary of the risks and benefits if available;
- Unmet medical need description that can be addressed by the IND; and
- The preliminary clinical evidence including: the conditions, outcome assessment, and patient monitoring; trials design, and analyses.

In case of submission with IND amendment, the cover letter should specify that the submission contains a RMAT request. The request should be in bold, uppercase letters as follows: **REQUEST FOR REGENERATIVE MEDICINE ADVANCED THERAPY DESIGNATION.**

If the request is submitted with an initial IND, the cover letter should specify that the submission contains both an initial IND and an RMAT. The request should be in bold uppercase letters as follows: **INITIAL INVESTIGATIONAL NEW DRUG SUBMISSION and REQUEST FOR REGENERATIVE MEDICINE ADVANCED THERAPY DESIGNATION.**

CBER will notify the sponsor of its decision no later than 60 calendar days after receipt request. In the case of a negative decision, CBER will include a written description of the rationale.

CBER may rescind the designation in case the product does not meet the criteria later in the development stages.

3.2.3 Post-approval requirements

The post-approval requirements for RMATs receiving accelerated approval may be satisfied by the following:

- The submission of clinical evidence, clinical studies, patient registries, or other sources of real-world evidence such as electronic health records;

- The collection of larger confirmatory data sets as agreed upon during product development; or
- Post-approval monitoring of all patients treated with such therapy prior to approval of the therapy.

Upon review of a BLA, CBER determines the type(s) of post-approval requirements (e.g., confirmatory clinical trials, patient registries, electronic health records, or other data collections) necessary to confirm the clinical benefits of an RMAT with accelerated approval.

The FDA may withdraw such marketing approval of a regenerative medicine therapy, including an RMAT, if the sponsor fails to comply with the requirements described in section 506(c) of the FD&C Act and 21 CFR 601.43(a).

3.2.4 *Interactions between sponsors and CBER review staff*

CBER recommends early discussions between sponsors of regenerative medicine therapies with the Office of Tissues and Advanced Therapies (OTAT). The "Formal Meetings Between the FDA and Sponsors or Applicants of PDUFA Products; Draft Guidance for Industry" describes standardized procedures for requesting, preparing, scheduling, conducting, and documenting formal meetings between sponsors and the FDA.

The Type B meetings described, include the pre-IND, end-of-phase II or pre-phase III, and pre-BLA meetings, represent critical points in the product development life cycle.

For some RMATs, it may be necessary for OTAT to engage in consultative or collaborative review with staff from other CBER offices or other FDA Centers. OTAT will work closely with the appropriate CBER office or FDA Center to reach a consensus regulatory decision (Table 3.6).

3.3 *Conclusion*

Regulatory pathways are evolving to ensure early approval of transformative therapies, and there are high expectations that they will bring hope for patients. We are still in an experimental phase for these new therapies, and regulatory requirements and processes will obviously adapt to future knowledge and accumulated regulators' experience. However, after obtaining a marketing authorization, ATMPs MA holders are struggling to achieve successful ATMPs reimbursement and market access. Payers are facing an immense pressure to make decisions based on limited evidence with high uncertainty, they are pushed by patients, healthcare professionals, and politicians to open access to these products.

Table 3.6 Summary of accelerated approval pathways benefits

	Benefits
Fast track	For drugs for serious conditions that fill an unmet medical need based on nonclinical or clinical evidence: • Frequent interactions with FDA, rolling review • Priority review if eligibility supported by clinical data submitted
Breakthrough therapies	Breakthrough requires a higher level of evidence than fast track: it expedites review of drugs for serious conditions which demonstrate substantial improvement on clinically significant endpoint(s) over available therapy based on preliminary clinical evidence Fast Track benefits, plus: • Intensive FDA guidance on efficient drug development • Organizational commitment to involve senior management in facilitating the development program
Accelerated approval	Only for products that generally provide a meaningful additional benefit over existing treatments. It applies to settings where disease course is long, and extended period would be required to measure the intended clinical benefit It enables approval of drugs for serious conditions that fill an unmet medical need based on surrogate endpoint likely to predict clinical benefit, or on clinical endpoint that can be measured earlier than irreversible morbidity or mortality
Priority review	Granted to products intended to treat serious conditions, and, if approved, represent a significant improvement in safety or effectiveness of the treatment of the condition. It may be granted to products with fast track, breakthrough, or RMAT designation • FDA take action within six months of receipt of application
RMAT	For regenerative medicines that treat, modify, reverse, or cure a serious or life-threatening disease or condition, based on preliminary clinical evidence Breakthrough Therapy benefits, plus: • Early interactions with FDA • Potential ways to support accelerated approval and satisfy post-approval requirements

Source: FDA, Guidance for industry expedited programs for serious conditions – drugs and biologics, 2014.

References

1. Regulation (EC) No 1394/2007 of the European parliament and of the council of November 13, 2007, on advanced therapy medicinal products and amending directive 2001/83/EC and regulation (EC) no 726/2004. 2007.
2. EMA. Summaries of scientific recommendations on classification of advanced therapy medicinal products [April 4, 2019]. Available from: https://www.ema.europa.eu/en/human-regulatory/marketing-authorisation/advanced-therapies/advanced-therapy-classification/summaries-scientific-recommendations-classification-advanced-therapy-medicinal-products.
3. Luria X, S. B. Handbook about Regulatory Guidelines and Procedures for the Preclinical and Clinical Stages of Advanced Therapy Medicinal Products (ATMPs).
4. The European Parliament and the Council of the European Union, Directive 2001/83/EC. 2001.
5. Kooijman M, van Meer PJ, Gispen-de Wied CC, Moors EH, Hekkert MP, Schellekens H. The risk-based approach to ATMP development—Generally accepted by regulators but infrequently used by companies. *Regulatory Toxicology and Pharmacology: RTP*. 2013;67(2):221–5.
6. EMA. Guideline on the risk-based approach according to annex I, part IV of Directive 2001/83/EC applied to Advanced therapy medicinal products. 2013. Available from: https://www.ema.europa.eu/en/documents/scientific-guideline/guideline-risk-based-approach-according-annex-i-part-iv-directive-2001/83/ec-applied-advanced-therapy-medicinal-products_en.pdf.
7. EMA. Guideline on the quality, non-clinical and clinical aspects of gene therapy medicinal products. 2018. Available from: https://www.ema.europa.eu/en/documents/scientific-guideline/guideline-quality-non-clinical-clinical-aspects-gene-therapy-medicinal-products_en.pdf.
8. EMA. Guideline on human cell-based medicinal products. 2008. Available from: https://www.ema.europa.eu/en/documents/scientific-guideline/guideline-human-cell-based-medicinal-products_en.pdf.
9. EMA. Guidelines relevant for advanced therapy medicinal products. Available from: https://www.ema.europa.eu/en/human-regulatory/research-development/advanced-therapies/guidelines-relevant-advanced-therapy-medicinal-products.
10. EMA. Procedural advice on the evaluation of advanced therapy medicinal product in accordance with Article 8 Of Regulation (EC) NO 1394/2007. 2009.
11. EMA. Explanatory note on general fees payable to the European Medicines Agency. 2016 [accessed on April 26, 2019]. Available from: http://www.ema.europa.eu/docs/en_GB/document_library/Other/2016/03/WC500203971.pdf.
12. EMA. European Medicines Agency pre-authorisation procedural advice for users of the centralised procedure. 2017 [accessed on April 26, 2019]. Available from: http://www.ema.europa.eu/docs/en_GB/document_library/Regulatory_and_procedural_guideline/2009/10/WC500004069.pdf.
13. Directive 2004/23/EC of the European Parliament and of the Council of March 31, 2004 on setting standards of quality and safety for the donation, procurement, testing, processing, preservation, storage and distribution of human tissues and cells. 2004.

14. EMA. Development support and regulatory tools for early access to medicines. 2016. Available from: https://www.ema.europa.eu/en/documents/other/development-support-regulatory-tools-early-access-medicines_en.pdf.
15. EMA. User guide for micro, small and medium-sized enterprises on the administrative and procedural aspects of the provisions laid down in Regulation (EC) No 726/2004 that are of particular relevance to SMEs. 2016.
16. EMA. Guideline on safety and efficacy follow-up—Risk management of Advanced Therapy Medicinal Products. 2008. Available from: https://www.ema.europa.eu/en/guideline-safety-efficacy-follow-risk-management-advanced-therapy-medicinal-products.
17. EMA. Risk Management of Advanced Therapies. 2009. Available from: http://www.ema.europa.eu/docs/en_GB/document_library/Presentation/2009/11/WC500007629.pdf.
18. FDA. Guidance for industry expedited programs for serious conditions–drugs and biologics. 2014. Available from: https://www.federalregister.gov/documents/2014/05/30/2014-12534/guidance-for-industry-on-expedited-programs-for-serious-conditions-drugs-and-biologics-availability.

chapter four

The need for new HTA reference case for cell and gene therapies

Up until early 2019, HTA bodies and payers seemed reluctant to approve reimbursement of cell and gene therapies based on the evidence available at time of launch. They were scrutinizing the value of the new innovative products, and tended to not recommend the products for reimbursement or postpone their final decisions mainly because of lack of long-term data. Cell and gene therapies could benefit from special processes for orphan drugs, cost-effectiveness thresholds, or exemptions for orphan drugs; however, these pathways do not offer a long-term solution for HTA challenges.

4.1 EU Reimbursement: current HTA frameworks in Europe

Despite their specificities, complexity, and potential benefits, cell and gene therapies are not considered different from other conventional therapies from HTA bodies' perspective. Some HTA bodies do not find a convincing rationale for adopting special assessment pathways for cell and gene therapies. HTA requirements are the same as other pharmaceuticals in both inpatient and outpatient settings, with no special pathways in the majority of the countries [1–3]:

- The HTA decisions in Germany are driven by clinical evidence. Germany is not open for outcome modeling and extrapolation of benefit. Therefore, the likelihood of added benefit acknowledgment may be limited when the follow-up is limited to one or a few years.
- The United Kingdom (UK) and Sweden are more open for those methodologies, but they always associate these long-term extrapolation analyses to high uncertainty.
- Spain and Italy are between the two groups. The two countries are less evidence driven but more driven by common sense and deliberative process, taking into consideration the drug price and budget impact.

An overview of EU HTA decision frameworks is provided in Table 4.1.

Table 4.1 Overview of HTA decision frameworks in some European countries

			Value Judgment		
HTA Agency	HTA Method	HTA Perspective (economic analysis)	Clinical Benefit	Cost-Effectiveness analysis	Budget Impact
France: TC, CEESP	Clinical data evaluation + economic evaluation	Payer (collective perspective)	High	High[a,b]	High[b]
Germany: IQWiG, G-BA	Clinical data evaluation	Payer (only drug budget impact)	High	Low[b]	Low
Italy: AIFA, regions	Clinical data evaluation + economic evaluation	Payer	High	Low	High/Moderate
Spain: SGCMPS, regions	Clinical data evaluation + economic evaluation	Payer	High	Low	High
Sweden: TLV	Economic evaluation	Societal	High/ Moderate	High[a]	Low/High[c]
England: NICE, Scotland: SMC	Economic evaluation	National health insurance	High	High	Moderate

TC: transparency committee, CEESP: Comité évaluation économique et santé publique, IQWIG: Institute of quality and efficiency in healthcare, GBA: Federal joint committee, AIFA: Italian medicines agency, SGCMPS: General sub directorate of quality of medicines and medical devices, TLV: Dental and pharmaceuticals benefits agency, NICE: National institute for health and care excellence, SMC: Scottish medicines consortium

[a] No formal threshold.
[b] only in certain cases/products.
[c] Low at national level and high for county councils.

ATMPs may benefit from pathways for orphan drugs or ultra-orphan drugs if they meet the qualifications criteria:

- **Germany:** Orphan drugs with projected annual sales to the GKV of less than €50 million are legally assumed to have an additional benefit. Thus, for these drugs, only the extent of the additional benefit is assessed.
- **NICE:** highly specialized therapies route (HST) for ultra-orphan drugs (prevalence <1 in 50,000 in England)
- **SMC:** Ultra-orphan drugs pathway—these are drugs with a prevalence <1 in 50,000 in Scotland. If the ultra-orphan drug is considered clinically effective by SMC, it will be made available on the NHS for at least three years while information on its effectiveness is gathered. The SMC will then review the evidence and may make a final decision on its routine use in NHS Scotland.

4.2 Assessment of current framework applicability to ATMPs in England: NICE mock appraisal

Some HTA agencies have started to consider whether current HTA models are appropriate for the evaluation of these potentially curative therapies. NICE has undertaken a mock appraisal to assess whether existing value assessment framework is suitable for assessing the value of regenerative medicines, including gene therapies [4]. NICE objectives were:

- to test the application of NICE appraisal methodology to regenerative medicines and identify the challenges and need adaptation of methodology
- to identify specific issues related to the appraisal of regenerative medicines using the current NICE appraisal process and decision framework; and
- to develop a framework for developing regenerative medicines to facilitate understanding of NICE assessment.

4.2.1 Hypothesis: benefits and costs

- For this exercise, a hypothetical gene therapy was considered: Hypothetical CAR T-cell specific to the antigen CD19 (Table 4.2), assumed to treat relapsed (two relapses or more) or refractory B-cell acute lymphoblastic leukemia (B-ALL).

Table 4.2 NICE mock appraisal hypothesis

	Scenario 1: Bridge to HSCT	Scenario 2: Curative Intent
TPPs	CAR T-cell induces short-term remission of disease in order to maximize the opportunity for successful HSCT.	CAR T-cell treatment is long-term remission/cure of disease (with or without HSCT).
Assumed individual patient level incremental QALY gain	7.46	10.07
Assumed price (acquisition cost)	£356,100	£528,660

Source: Hettle, R. et al, *Health Technol. Assess.* (*Winchester, England*), 21, 1–204, 2017.

The product has the potential for profound patient benefits, but there is also very high uncertainty around the actual benefits that these products would deliver. In analyzing and presenting uncertainty, the York team determined:

- Incremental cost-effectiveness ratio (ICER) and probability of being cost effective at the applicable threshold.
- Incremental Net Health Effect (NHE): Considering the impact of recommending the therapy on population-level health.
- Consequences of decision uncertainty: This reflects the potential magnitude of NHE that could be gained if uncertainty surrounding potential decisions could be resolved.

NHE and decision uncertainty values were based on 38 patients per year over a 10-year period.

4.2.1.1 Summary of results

- **Scenario 1 Results: Bridge to HSCT**
  In this example, the key outcome from therapy is clinical remission to allow HSCT. It could be estimated with reasonable accuracy from even the minimum evidence set.
  - Increasing maturity of evidence had relatively low impact on the consequences of decision uncertainty.

- **Scenario 2 Results: Curative intent**
  - Increasing maturity of evidence had a major impact on reducing uncertainty.
  - The additional evidence increases the certainty around the curative benefits and cost- effectiveness of the treatment.
- **Conclusion**
  The relative treatment benefits generated from single-arm trials are likely to be optimistic unless the historical data have accurately estimated the efficacy of the control agent. Surrogate endpoints, on average, overestimate treatment effects.

NICE concluded that:

> Evidential challenges were **not unique to regenerative medicines** or not beyond the scope of existing assessment frameworks.
>
> - The existing appraisal methods are **applicable** to cell and gene therapies.
> - **Quantification of decision uncertainty** was key in decision-making.
> - **Innovative payment mechanisms** may play an important role where uncertainty is substantial.

4.3 Review of HTA experience in Europe

HTA reports available at time of writing were reviewed to identify the main limitations criticized by HTA bodies in EU5. Table 4.3 summarizes the HTA recommendations for ATMPs available so far (December 2019). England and Italy were the countries where most ATMPs were reimbursed, and Spain showed the longest delays in assessing ATMPs.

4.3.1 France

4.3.1.1 Glybera®

Glybera® was not recommended by TC in November 2015. TC considered that the actual benefit (AB) is insufficient, its benefit could not be established based on the clinical trials design (open label), small sample size and the absence of a sustained effect beyond one year.

Table 4.3 ATMPs HTA review in EU5

	France	Germany	England	Scotland	Italy	Spain
Glybera® (withdrawn MA)	Not recommended AB: insufficient	Non-quantifiable added benefit	NA	NA	NA	NA
Imlygic®	NA	No added benefit	Recommended with restriction of indication	NA	List Cnn	Not recommended
Strimvelis®	NA	NA	Recommended	NA	Recommended	NA
Kymriah®	Recommended AB: important IAB: III/IV[a]	Non-quantifiable added benefit	Funded via CDF for ALL	Recommended	Recommended	Recommended
Yescarta®	Recommended AB: important IAB: III	Non-quantifiable added benefit	Funded via CDF	Recommended	Recommended	Recommended
Luxturna®	Recommended IAB II	Considerable added benefit	Recommended	Ongoing assessment	NA	NA
Provenge® (withdrawn MA)	NA	Non-quantifiable added benefit	Not recommended	NA	NA	NA

(Continued)

Table 4.3 (Continued) ATMPs HTA review in EU5

	France	Germany	England	Scotland	Italy	Spain
Zalmoxis® (withdraw)	Not recommended	Non-quantifiable added benefit	NA	NA	Recommended	NA
Alofisel®	Recommended with restriction of indication (IAB IV)	Non-quantifiable added benefit	Not recommended	Not recommended	NA	NA
Chondrocelect® (withdrawn MA)	Not recommended AB: insufficient	Not eligible to early benefit assessment	No longer included in the appraisal of ACI	NA	NA	Recommended
MACI® (withdrawn MA)	NA			NA	NA	NA
Holoclar®	Recommended with restriction of indication IAB: IV		Recommended with restriction of indication	NA	Recommended	NA
Spherox®	NA	NA	Recommended	NA	NA	NA

NA: not assessed or no published decision, CDF: Cancer Drugs Fund; ACI: autologous chondrocytes implementation; ALL: Acute lymphoblastic leukemia; Cnn list: not yet assessed, H list: hospital only.

Yescarta®

[a] IAB III in ALL and IAB IV in DLBCL.

Limitations of clinical data criticized by TC were as follows:

- Clinical relevance of primary endpoint (reduction in the triglyceride level) is debatable: surrogate endpoint.
- Weak clinical trial methodology exists (comparing before/after, open label).
- The sample size was small (eight patients) compared to the number of available patients.
- Moderate effect on triglycerides and on episodes of pancreatitis has been observed, but this effect was not sustained in the medium and long term (return to the baseline triglyceride level one year after the injection).
- There was inter-patient heterogeneity of the treatment response.
- A lack of information regarding the diet during the trial.
- Post-hoc incidence and severity of pancreatitis analysis cannot allow to determine the impact of Glybera® on pancreatitis.
- Transferability to real life is not completely guaranteed especially due to the small sample size that makes the generalizability of results difficult.

HAS highlighted the high uncertainty in the clinical evidence:

- There was uncertainty about the short- and long-term tolerance of the gene therapy.
- No long-term efficacy was detected; the effect was not sustained in the medium and long term.
- It was impossible to quantify the impact of Glybera® on morbi-mortality.
- There was poor methodology of the studies (open label, noncomparative design, comparing before/after), small sample size, and clinical relevance of the efficacy endpoint.
- The data on the incidence and severity of the episodes of pancreatitis do not provide proof that Glybera® has an impact in the prevention of pancreatitis.

4.3.1.2 Chondrocelect®

Chondrocelect® was assessed twice by the TC, and in both conclusions the TC did not recommend Chondrocelect® for reimbursement. It had an insufficient actual benefit due to the lack of efficacy demonstrated using clinical criteria, observed adverse effects which, although not serious, are more common than with the microfracture technique, and may potentially jeopardize the benefit of this major surgery and have a potentially adverse effect on patients' rehabilitation. In addition, the technical complexity necessitates two hospitalizations—one for arthroscopy to harvest the chondrocytes, the other for arthrotomy for their implantation and their maintenance by means of a periosteal flap, the suture for which on the periphery of the cartilage is tricky and may not last. Therefore, the efficacy/adverse effects ratio for Chondrocelect® has not been clearly established [5].

4.3.1.3 Holoclar®

The actual benefit was considered important in the treatment of moderate to severe limbal stem cell deficiencies fulfilling the following criteria:

- Superficial cornea neovascularization in at least two quadrants in one of the two eyes
- Central cornea problem
- Altered visual acuity

The actual benefit was considered insufficient for other conditions not fulfilling the criteria above. The improvement in actual benefit (IAB) was level IV [6].

4.3.1.4 Zalmoxis®

Zalmoxis® was considered to have insufficient actual benefit to justify its reimbursement for adjuvant treatment during haploidentical hematopoietic stem cell transplantation in adults presenting with high-risk malignant blood disorders. It has limited efficacy data and failed any comparative data relative to the reference strategy for haploidentical transplantation.

4.3.1.5 Yescarta®

TC recommended Yescarta® in the indications of the label; it was given an important actual benefit and incremental actual benefit level III. TC considered that the submitted data prove Yescarta's short-term efficacy on the overall response (around 50% of ITT population) and on the overall survival in diseases with high medically unmet needs. TC recognized that there are uncertainties around the effect size and long-term efficacy. In addition, Yescarta® has significant toxicity in the short term with lack of safety results in the long term [7].

Yescarta® will be assessed annually by the transparency committee based on long-term clinical data and real-world evidence from the French registry. The time of production will also be assessed.

4.3.1.6 Kymriah®

TC recommended the second CAR-T cell, Kymriah®, for reimbursement in France. The actual benefit was important for both label indications, and the IAB was III for ALL and IV for DLBCL.

ALL: The efficacy data have shown important remission after three months (around 67% of ITT population) maintained in 40% of the patients after a median follow-up of nine months. There is high uncertainty about the effect size due to lack of direct comparison versus the standard of care in addition to uncertainty about long-term efficacy. The toxicity in the short term is significant, and there is no data on long-term safety [8].

DLBCL: The efficacy data showed limited efficacy in the short term on complete response (24% on ITT population) and on the overall survival with a median follow-up of seven months. There is uncertainty about the effect size due to the lack of direct comparison versus the standard of care and no long-term efficacy data. The toxicity in the short term is significant, and there is no data on long-term safety [9].

Like Yescarta®, Kymriah® will be assessed annually by the transparency committee based on long-term clinical data and real-world evidence from the French registry. The time of production will also be assessed.

4.3.1.7 Luxturna®

Luxturna® was considered to have high clinical benefit in inherited retinal dystrophy caused by biallelic RPE65 mutations and substantial clinical added value in the therapeutic strategy: important actual benefit and IAB level II. It will be reassessed after five years. It significantly improved the functional vision of patients, assessed via multi-luminance mobility testing (MLMT), one year after treatment of both eyes. The decision to initiate treatment should be the subject of a multidisciplinary team meeting and be based on an array of tests, in particular to determine a sufficient number of viable cells.

4.3.2 Germany

4.3.2.1 Glybera®

The German Federal Joint Committee (G-BA) assessed Glybera® and granted it "unquantifiable additional benefit." Then, the positioning was changed to a hospital-only product; therefore, direct price negotiations between hospitals and payers are required. Before its withdrawal, one patient received Glybera® in Germany at Charité in Berlin in September 2015 [10].

- **Limitations of clinical data:**
  - The outcome "pancreatitis" was used post hoc to evaluate the efficacy of the intervention studies and is poorly operationalized.
  - Due to the lack of operationalization of pancreatitis in the intervention studies, this outcome is not considered valid. The data

reviews show that it was initially planned to record all pancreatitis, including those that did not lead to hospitalization. Instead, only hospitalization-related pancreatitis events and abdominal pain were included, making it possible to underestimate the events before and after the study medication, which also jeopardizes the validity of the outcome.
    - Change in Chylomicron values has not been validated as a surrogate parameter for pancreatitis.

- **Uncertainty:**
    - Due to the rareness of the disease and the absence of resilient data, there has to be a high insecurity concerning the indications for the target population.
    - It remains unclear how many patients with pancreatitis really suffer from severe or multiple pancreatitis. Thus, the company's estimation rather reflects an upper limit.
    - There was insufficient data on safety and efficacy; quantification of added benefit was not possible.
    - Results on pancreatitis must be considered uncertain and biased due to noncomparative trials.
    - Hospitalizations were not systematically recorded as separate endpoints.
    - TG-lowering effect could not be detected in the long-term data after one year at all.
    - A possible underestimation of number of pancreatitis events occurred; only hospitalization-related pancreatitis events and abdominal pain were included.

4.3.2.2 *Provenge®*

G-BA concluded that the clinical added benefit of Provenge® is "non-quantifiable" due to bias of the treatment effect of sipuleucel-T: influence from concomitant treatments for the effect estimates of all studies cannot be excluded with certainty. Lower mortality cannot be explained solely by the effect of sipuleucel-T. Advantages in overall survival are accompanied by negative effects in the form of side effects [11].

4.3.2.3 *Imlygic®*

The German health technology assessment body (IQWiG) concluded that the manufacturer dossier contained no data suitable for assessment. G-BA defined three comparators based on the treatment previously taken by the patients. These comparators were not used in the trials [12].

The clinical trial design was versus GM-CSF, which was not considered the appropriate comparator for the three subgroups identified by G-BA:

- Treatment-naive adults with BRAF V600 mutant tumor,
- Treatment-naive adults with BRAF V600 wild-type tumor,
- Pretreated adults.

4.3.2.4 *Zalmoxis®*

Zalmoxis® was assessed with "non-quantifiable" extent of added benefit. G-BA considered that the pivotal study TK007—a one-arm, uncontrolled Phase I/II study—as well as individual patient data from the ongoing Phase III study TK008 and a historical comparison with data from a registry were not sufficient to provide valid assessment of the added benefit.

G-BA criticized that the subpopulation accounted for just under 12% (15 out of 127) of patients included in the intervention arm of the study. In addition, since no characteristics of the 15 subpopulations were reported, results for patients of the control arm were not available and it was not a pre-planned evaluation. Therefore, these data were not suitable for the assessment of the extent of added benefit of Zalmoxis®.

4.3.2.5 *Alofisel®*

Alofisel® was assessed with "non-quantifiable" extent of added benefit. G-BA was unable to decide on the additional benefit due to high uncertainty. The overall clinical evidence was considered limited. The results are subject to strong uncertainties due to existing limitations in the assessment of clinical remission (clinical examination by gentle finger compression: subjective examination procedure without standardization, overevaluation and underevaluation of fistula closure, internal branching and the more proximal sections of the fistula cannot be assessed).

4.3.2.6 *Kymriah®*

DLBCL: Kymriah® is licensed as an orphan drug. An added benefit is provided for by law.

The pharmaceutical company provided the study results of the single-arm Phase II study JULIET, the supportive case series by Schuster et al. [13] as well as non-adjusted indirect comparisons to various historical control studies.

The case series by Schuster et al. is not used for the benefit assessment because, among other things, relevant information on the study methodology and study characteristics is missing.

There is a lack of data for the pivotal single-arm study JULIET. For the more recent data cuts, there is a lack of necessary information on the course of the study and the conduct of the study, so that these cannot be used for the benefit assessment. The data from March 8, 2017, which can be used for the benefit assessment, are subject to great uncertainty due to the very short median follow-up time and the incomplete recruitment of patients. In addition, there are no overall survival analyses for the ITT population for this data cut. Leukapheresis, waiting time until availability of the product, and the frequently associated administration of bridge chemotherapy as well as lymphocyte-depleting chemotherapy are inherent components of treatment with tisagenlecleucel. The influence of these components on the treatment of patients with tisagenlecleucel in the clinical care context can only be adequately mapped by considering the ITT population. Overall, no reliable conclusions can be drawn on the extent of the added benefit due to the lack of data.

On the basis of the indirect historical comparisons presented, no sufficiently valid conclusions on the extent of the added benefit of tisagenlecleucel can be derived at present on the basis of the insufficient data available from the JULIET study and further uncertainties regarding comparability of the studies used for the indirect comparison.

Overall, a nonquantifiable added benefit is found only from a legal point of view.

The decision expired on March 15, 2020.

ALL: Kymriah® has orphan drug approval. An additional benefit is provided for by law.

The pharmaceutical company presents the study results of the single-arm Phase II studies ELIANA and ENSIGN, the single-arm Phase I/II study PEDICAR, and non-adjusted indirect comparisons to various historical control studies.

The PEDICAR study is not included in the benefit assessment, as among other things, a risk of bias of study results due to multiple administration of tisagenlecleucel cannot be excluded.

Data for the pivotal single-arm study ELIANA and the supportive single-arm study ENSIGN are insufficient. For the more recent data cuts, there is a lack of necessary information on the course of the study and the conduct of the study, so that these cannot be used for the benefit assessment. With regard to the data cuts that can be used for the benefit assessment, the data are subject to considerable uncertainty due to the very short median follow-up time and the incomplete recruitment of patients. In addition, no patient characteristics and overall survival assessments are available for these ITT population data cuts. Leukapheresis, the waiting time until availability of the product, and the frequently associated administration of bridge chemotherapy as well as lymphocyte-depleting chemotherapy are inherent components of treatment with tisagenlecleucel. The influence of these components on the treatment of patients with tisagenlecleucel in the clinical care context can only be adequately mapped by considering the ITT population.

Overall, no reliable conclusions can be drawn on the extent of the added benefit due to the lack of data.

On the basis of the indirect historical comparisons presented, no sufficiently valid conclusions on the extent of the added benefit of tisagenlecleucleus can currently be derived due to the insufficient data available from the ELIANA and ENSIGN studies and further uncertainties regarding comparability with the studies used for the indirect comparison.

In the overall view, a nonquantifiable added benefit is determined solely from a legal point of view.

The decision is valid until March 15, 2020.

4.3.2.7 Yescarta®

Yescarta® is an orphan drug. Consequently, an added medical benefit is proven by the regulatory approval. G-BA determines the extent of the added benefit for the number of patients and patient groups for whom there is a therapeutically relevant added benefit.

The G-BA derives an unquantifiable added benefit from the mortality data of the pivotal single-arm Phase I/II study ZUMA-1 and the indirect historical comparison with the retrospective study SCHOLAR-1.

Due to the indirect historical comparison and other relevant uncertainties, a valid quantification of the extent of the effect on overall survival is not possible.

The single-arm study design of the ZUMA-1 study does not allow a comparative evaluation of the other endpoints of morbidity and side effects.

The quality of life of patients was not assessed in the ZUMA-1 study.

The patient population is given as 440 to 700.

Strict, detailed guidelines for the application of the product apply to patients, the medical staff involved, and the medical facilities.

Due to the advanced stage and poor prognosis of the further course of the disease, high weight was attached to the comparative results for overall survival in the overall evaluation.

The overall G-BA rating is nonquantifiable added benefit.

This decision is time-limited to May 15, 2022.

4.3.2.8 Luxturna®

Luxturna® was granted a considerable added benefit by G-BA.

The decision is driven mainly by statistically significant, clinically relevant results on MLMT, FST, and perimetry.

- Statistically significant change in visual acuity could not be demonstrated for Neparvovec compared to watchful waiting. There are no suitable data for the quality of life for the benefit assessment.
- There was uncertainty about long-term benefits.

Limitations:

- The significance of the two submitted studies was classified as limited, justified by the respective study design as well as the existing uncertainties in relation to the subjective tests used to record the morbidity and their repeatability. On the basis of the available studies, no reliable statements could be made on the sustainability of the positive effects. Overall, there was a hint of significant added benefit of Neparvovec over observational waiting for the treatment of adult and pediatric patients with vision loss due to hereditary retinal dystrophy, which was based on proven biallelic RPE65 mutations, and who have sufficiently viable retinal cells the resolution is limited to December 31, 2021, due to outstanding study results.
- The evaluations presented in the context of this evaluation procedure did not allow a conclusive assessment of the added benefit for Voretigen Neparvovec with sufficient certainty, especially due to the lack of long-term data on all patient-relevant outcomes. Without long-term data, the sustainability of the positive effect of gene therapy cannot be assessed. The long-term effects of gene therapy with Luxturna, in particular with regard to the safety profile, are currently the subject of further investigations within the framework of the approval requirements.
- To assess the sustainability of the effects in the dossier, long-term data on patient-relevant outcomes—the interim analysis of extension study 302 for all patient-relevant outcomes and the safety data of the registry—should be provided for the reassessment of the benefit after the deadline. The G-BA considered that a deadline of December 31, 2021 was appropriate.

4.3.3 England

4.3.3.1 Provenge®

In January 2015, NICE published the final appraisal for Provenge® stating: "Sipuleucel-T is not recommended within its marketing authorization for treating adults who have asymptomatic or minimally symptomatic metastatic non-visceral hormone-relapsed prostate cancer for which chemotherapy is not yet clinically indicated." NICE considered that Provenge® was not cost-effective and did not meet the criteria for end-of-life consideration. The committee had uncertainties regarding the indirect comparison with arbiterone. In addition, NICE concluded that the trials did not show that sipuleucel-T delayed disease progression compared with a placebo [14].

- The three trials included patients who had previously received chemotherapy, although the MA indication was for people with metastatic hormone-relapsed prostate cancer for which chemotherapy is not yet clinically indicated.
- None of the trials showed that sipuleucel-T prolonged time to disease progression.
- There was insufficient evidence to establish whether the clinical effectiveness of sipuleucel-T was different in the low-PSA subgroup compared with the rest of the population.
- There was uncertainty surrounding the results of the indirect comparison.
- Uncertainty about the estimates of cost-effectiveness existed: inadequate model structure.
- The Committee noted that it had not been presented with evidence on the effectiveness of sipuleucel-T at different places in the treatment pathway, and it was aware that the marketing authorizations for sipuleucel-T and abiraterone are almost identical. Therefore, the Committee concluded that it could only appraise sipuleucel-T based on the evidence presented and in line with the marketing authorization.

4.3.3.2 Imlygic®

NICE considered that Imlygic® was not cost-effective. However, after providing more evidence from the manufacturer and agreeing on a straight confidential discount, NICE approved the use of Imlygic® for patients for whom treatment with systemically administered immunotherapies is not suitable [15].

- **Limitations of clinical data**
  - The comparator used in the trial GM-CSF is equivalent to placebo and not used in practice.
  - There was potential for bias in the trial because of limited blinding, differences in the withdrawal rates in the two arms, and the use of a non-validated primary endpoint.
  - Post-hoc analysis of a subgroup was conducted.
  - The right comparator from NICE perspective is ipilimumab.
  - There were substantial differences in patient characteristics between talimogene laherparepvec and ipilimumab trials in the indirect comparison.

- **Limitations of economic model**
  - There was a lack of suitable effectiveness inputs in the economic model.
  - A determination of ICER versus Ipilimumab could not be made.

- Imlygic® was considered cost-effective only in the patient population for whom treatment with systemically administered immunotherapies is not suitable: £23,900 (versus decarbazine) £24,100 (versus best supportive care).

4.3.3.3 *Holoclar®*

In August 2017, NICE recommended Holoclar® for the treatment of moderate to severe limbal stem cell deficiency after eye burns; it should be used to treat one eye only after a conjunctival limbal autograft, or in case of insufficient tissue for a conjunctival limbal autograft, or it is contraindicated. A discount was agreed with the company.

The treatment of both eyes is recommended for research only and when there is not enough tissue for a conjunctival limbal autograft.

> Moderate to severe limbal stem cell deficiency is defined by the presence of superficial corneal neovascularisation in at least two corneal quadrants, with central corneal involvement, and severely impaired visual acuity. [16]

4.3.3.4 *Strimvelis®*

Strimvelis® was recommended by NICE; it was assessed via highly specialized technologies route (HST). The cost of Strimvelis® was considered high (594,000€) and there were some uncertainties in the evidence. However, it was considered that Strimvelis® is likely to provide important benefits for people with ADA–SCID, at a cost that provides value for money in the context of a highly specialized service [17].

4.3.3.5 *Kymriah®*

ALL: Kymriah® was recommended for use within the Cancer Drugs Fund (CDF) as an option for treating relapsed or refractory ALL in people aged up to 25 years, only if the conditions in the patient access scheme (confidential discount) are followed. The evidence on efficacy was uncertain, and the evidence to determine the costs of treating side effects and whether people will need a subsequent stem cell transplant was insufficient. Collecting more data on overall survival, subsequent stem cell transplant rates, and immunoglobulin usage will reduce the uncertainty in the clinical- and cost-effectiveness evidence. Therefore, it was recommended for use in the CDF to collect this data in the real world [18].

DLBCL: Similarly, in February 2019, NICE recommended Kymriah® for adults with relapsed or refractory DLBCL to be funded through the CDF [19].

4.3.3.6 *Yescarta®*

Similarly, Yescarta® is recommended for use within the CDF as an option for treating relapsed or refractory DLBCL or PMLBCL in adults after two or more systemic therapies, only if the conditions in the managed access agreement are followed. Evidence from a small, single-arm study suggested that people having axicabtagene ciloleucel have clinically meaningful overall and progression-free survival. However, the evidence was uncertain because there was a limited follow-up and no direct data comparing axicabtagene ciloleucel with salvage chemotherapy. Limitations in the available data meant that the exact size of the benefit of axicabtagene ciloleucel compared with salvage chemotherapy was unknown. There was also not enough evidence to determine the costs of treating side effects.

Axicabtagene ciloleucel met NICE's criteria to be considered a life-extending treatment at the end of life. The most plausible cost-effectiveness estimates for axicabtagene ciloleucel compared with salvage chemotherapy were uncertain because of immature survival data. Collecting further data on progression-free survival, overall survival, and immunoglobulin usage will reduce the uncertainty in the evidence [20].

4.3.3.7 *Luxturna®*

Luxturna® was assessed via the HST pathway for ultra-orphan drugs. Clinical trial evidence showed that, in the short term, voretigene neparvovec improved vision and prevented the condition from getting worse. There was no long-term clinical evidence, but it was biologically plausible that the treatment effect was likely to continue for decades.

Some assumptions in the economic modeling were uncertain, particularly around the utility values and how long the treatment effect lasts. Despite the uncertainties, it was considered that voretigene neparvovec is likely to provide important clinical benefits for people with RPE65-mediated inherited retinal dystrophies.

There was a simple discount patient access scheme for voretigene neparvovec.

4.3.3.8 *Alofisel®*

It was not recommended by NICE due to several reasons:

- The evidence on the natural history of the disease and outcome of current practice in the UK was limited.
- Clinical-effectiveness data for darvadstrocel was from only one trial with a relatively short time frame.
- ADMIRE-CD showed a statistically significant benefit of darvadstrocel compared with placebo, but by one year more than 50% of patients with remission had relapsed.

- There was high uncertainty in the generalizability of the placebo arm to UK clinical practice; therefore, it was uncertain whether the benefit shown for darvadstrocel in the trial would translate to the same benefit compared with standard care in the NHS.

4.3.3.9 Spherox®

It was recommended as an option for treating symptomatic articular cartilage defects of the femoral condyle and patella of the knee in adults, only if:

- The person has not had previous surgery to repair articular cartilage defects,
- There is minimal osteoarthritic damage to the knee, and
- The defect is over 2 cm^2.

Despite some limitations in the economic evaluation, NICE considered that Spherox® was cost effective, and it was as effective in the short term as microfracture, which is the most commonly used surgical option. But little data is available beyond two years. It was considered to have greater benefit in articular cartilage defects larger than 2 cm^2.

4.3.4 Scotland

4.3.4.1 Yescarta®

In the initial assessment, Yescarta® was not recommended in Scotland although the criteria for end-of-life and ultra-orphan therapies were satisfied. There were no comparative data available, as only single-arm study was submitted. Data on patient-reported quality of life outcomes was very limited. ORR was assessed by investigators, the open-label study design brought about a potential for bias. A range of limitations existed in the economic analysis derived from the indirect comparison methods.

In the reassessment, SMC considered the benefits of axicabtagene ciloleucel in the context of the SMC decision modifiers that can be applied when encountering high cost-effectiveness ratios and agreed that as Yescarta® is an ultra-orphan medicine, SMC can accept greater uncertainty in the economic case.

4.3.4.2 Kymriah®-DLBCL

Kymriah® was not recommended in the initial assessment in Scotland because the cost in relation to treatment benefit was not sufficiently justified. In addition, SMC criticized other limitations including the economic analysis was based on a naïve indirect comparison, bridging chemotherapy might have an impact on the treatment effect, and the limitations with the modeling for the endpoint overall survival.

In the reassessment, SMC considered that the new-proposed Patient Access Scheme improved the cost-effectiveness; thus, Kymriah® was recommended for DLBCL treatment.

4.3.4.3 Kymriah®-B cell ALL

Kymriah® was recommended for the treatment of B-cell ALL by taking into account the benefit of PAS. SMC was concerned about the noncomparative study design, and a number of limitations associated with indirect comparison including the heterogeneity in the study population and follow-up duration. A high level of uncertainty could be accepted considering that B-cell ALL met the criteria for ultra-rare diseases.

4.3.5 Italy

Five ATMPs have been approved on list H: Strimvelis®, Zalmoxis®, and Holoclar®, Yescarta®, and Kymriah®.

Holoclar® (Determina AIFA n. 204/2017) and Strimvelis® (Determina AIFA n. 1028/2016) were reimbursed with a managed entry agreement: payment by results. Zalmoxis® was reimbursed with a flat cost per patient (Determina n. 139/2018).

Kymriah® (Determina n. 1264/2019) and Yescarta® (Determina n. DG/1643/2019) are innovative therapies, benefiting from an innovation fund.

4.3.6 Spain

Chondrocelect® was reimbursed in Spain before its MA withdrawal. Imlygic®, Yescarta®, and Kymriah® are reimbursed in Spain.

4.4 US reimbursement

The US health insurance system differs from the EU health system. It is made up of private insurance companies and public/government payers (Medicare and Medicare). There are no formal HTA frameworks in the US; there are general value frameworks, including assessment methods for oncology drugs. And like in the EU, there are no value frameworks that specifically address ATMPs.

The US' Institute for Clinical and Economic Review (ICER) is an independent institution established in 2006 to evaluate the clinical and economic value of new drugs and medical tests. ICER is known as the US' independent watchdog on drug pricing. It defines a "value-based price benchmark" reflecting how each drug should be priced to appropriately reflect long-term improved patient outcomes. It aims to facilitate discussions between payers and pharmaceutical industries [20].

The role of ICER in the US is growing and could eventually have an impact on reimbursement and pricing of new therapies.

Like some EU HTA bodies, one of the key tools used by ICER to assess a new therapy is the ICER. To determine the "value-based benchmark" for a drug, ICER collaborates with major stakeholders, including patients, healthcare providers, and manufacturers, to look at four key criteria [21]:

- efficacy,
- incremental benefit over available therapies/standard of care,
- cost savings,
- cost of treating the target population.

However, unlike EU HTA bodies, ICER has cautioned that existing QALY assessments may not be adequate for regenerative medicines. ICER is working with the Alliance for Regenerative Medicine to develop a value framework for regenerative medicines.

4.4.1 *ICER initiative: methods adaptation in the assessment of potential cures*

On August 6, 2019, ICER released a draft set of methods adaptations to its value framework to be applied in the assessment of potential cures and other treatments that qualify as "single or short-term transformative therapies," or SSTs. These proposals are intended to provide decision-makers with a reliable evaluation of SSTs' uncertainty, value, and value-based pricing. They do not fundamentally alter ICER's approach to value assessment. This report was finalized in November 2019 and is available online in the ICER website providing adapted value assessment methods for high-impact SSTs [22].

SSTs include two subcategories:

- Potential cures that can eradicate a disease or condition; and
- Transformative therapies that can produce sustained major health gains or halt the progression of significant illnesses.

Some of the key proposals include:

- To address uncertainty resulting from limitations in evidence using standardized multiple methods.
- Threshold analyses in order to identify the time horizons needed for assumptions about short-term efficacy to remain true, in order for curative therapies to reach standard cost-effectiveness thresholds.
- Mechanisms to use probabilistic sensitivity analysis to provide value-based pricing recommendations linked more directly to the degree of uncertainty in the data.

4.4.2 *Advanced therapies reimbursement in the US*

Current cell and gene therapies are administered typically in a hospital setting. Therefore, they are reimbursed through the medical benefit, not the pharmacy benefit. There will likely not be major involvement from a middleman, such as a pharmacy benefit manager (PBM).

US payers have not imposed major restrictions on Imlygic®, Kymriah®, Luxturna®, and Yescarta® beyond their approved FDA labels. Some payers have restricted their use to a subpopulation that they consider may benefit the most from the therapy. These are in line with the inclusion or exclusion criteria for the clinical trials.

4.4.3 *Review of CAR-T cell therapies case study: Kymriah® and Yescarta®*

4.4.3.1 *ICER assessment of CAR-T therapies*

In a 2018 report, ICER evaluated the comparative clinical effectiveness and value of Kymriah® and Yescarta® in two indications: B-cell ALL in children and young adults, and aggressive B-cell lymphoma in adults (both refractory or in second or greater relapse), versus chemotherapy.

Some limitations were highlighted, including the single-arm design, small number of patients in the trials, and short median follow-up time leading to uncertainty for estimates of clinical efficacy in the long term [24].

ICER determined that Kymriah® and Yescarta® were cost effective (total discounted costs) at $667,000 in B-cell ALL and $617,000 in DLBCL respectively.

For Kymriah®, the analysis identified discounted life years (LYs) gained of 10.34 and QALYs of 9.28. Kymriah® was compared with clofarabine, which had a $337,000 total discounted cost, 2.43 LYs, and 2.10 QALYs; as a result, the incremental cost-effectiveness ratio for Kymriah® versus clofarabine is $42,000 per LY gained and $46,000 per QALY.

While for Yescarta®, the analysis showed 7.35 LYs and 5.87 QALYs. It was compared with chemoimmunotherapy regimens used in Kite Pharma's SCHOLAR-1 trial (R-DHAP [rituximab, dexamethasone, high-dose ara-C cytarabine, Platinol]), which had a total discounted cost of $155,000, LYs of 3.23, and QALYs of 2.48. The incremental ratio of Yescarta® was $112,000 per LY and $136,000 per QALY [24].

4.4.3.2 *CMS coverage*

To date, coverage of CAR-T therapies varied because the decisions are made at the local Medicare contractor level.

CMS began a national Medicare coverage analysis for CAR-T in May 2018. The first draft of the analysis was published on February 15, 2019 [25]. In this draft, CMS proposes Coverage with Evidence Development for CAR

T-cell Therapy. The proposed decision would provide nationwide consistency in CMS' coverage of the innovative new cancer therapy, to improve patient access and ensure appropriate evidence generation. The proposed National Coverage Determination would require Medicare to cover the CAR-T at a national level when it is offered in a CMS-approved registry or clinical study, in which patients are monitored for at least two years post-treatment. Two years later, the evidence from the registries and studies would help CMS identify the types of patients that benefit from CAR T-cell therapy, informing a future decision by the agency regarding the types of cases in which Medicare would cover the treatment with no registry or trial requirement.

4.5 Challenges at HTA level

Cell and gene therapies' HTA review shows that one of the main challenges facing these therapies is at valuation level due to the lack of comparative trials, short-term data, and lack of long-term clinical data on safety and efficacy. These factors lead to high uncertainty, which constitutes the main hurdle facing cell and gene therapies. In addition to the high uncertainty, HTA bodies have a lack of flexibility in assessing these therapies.

4.5.1 Uncertainty

The uncertainty is mainly at two levels: uncertainty on efficacy, safety and long-term efficacy and the uncertainty on cost-effectiveness.

4.5.2 Uncertainty around short-term efficacy

4.5.2.1 No comparative trials

The gold standard for HTA bodies, especially the clinically driven HTA, e.g., Germany and France, is the head-to-head randomized clinical trials (RCTs). The comparator should be the standard of care in the specific country for an indication. But in some cell and gene cases, where comparative trials are not feasible, indirect comparisons are increasingly used. RCTs may not be feasible in situations where patient recruitment and ethical considerations present challenges, such as in small patient populations, high-risk patients.

Trial designs usually used for cell and gene therapies:

- **Single-arm:** In some cases, head-to-head trial design may not be possible due to unavailable appropriate comparator. Furthermore, randomized placebo controlled trials may not be feasible due to the limited number of patients or ethical concerns.
- **Open-label:** Due to the complex administration of some cell and gene therapies, double-blind trials may not always be possible.

- **Small sample size:** Small sample sizes are usually used due to the small target population of orphan and ultra-orphan diseases: Statistical significance can be achieved despite small sample size because of high benefit, but power may be low.
- **Short duration:** Duration of follow-up may sometimes be considered short for cell and gene therapies trials; long-term adverse reactions and sustainability of the benefit may raise high uncertainty.
- **Surrogate endpoints are used in most clinical trials:** Physical measurement of a specific outcome considered as a valid predictor (or representative) of the final result.

All these trial parameters are accepted by regulatory bodies and develop a level of evidence considered sufficient to establish a positive benefit-risk balance to grant marketing authorization. However, payers' requirements are more stringent.

4.5.3 *Uncertainty around long-term efficacy and safety*

Cell and gene therapies reaching the market via expedited regulatory pathways are likely to have short-term data at time of launch. In addition, direct long-term clinical data collection may be incompatible with clinical development timelines. Therefore, long-term clinical benefits are usually extrapolated from short-term trial data. It involves multiple parametric and non-parametric models that are validated through statistical considerations and clinical expert opinion on biological plausibility. NICE is the only HTA body that provides formal guidance on long-term data extrapolation.

Extrapolated data and indirect observations are considered weaker evidence than direct clinical data. However, in the US, the 21st Century Cures Act (Section 2.3.1) specifically allows for accelerated approval based on surrogate or intermediate clinical trial endpoints.

4.5.4 *Uncertainty around cost-effectiveness*

In cost-effectiveness modeling, the sources of uncertainties include the inputs, the methodology, and the structure. There are a number of sources of uncertainty within the available evidence at time of launch that are relevant to estimating the cost-effectiveness; for example, uncertainty in the treatment benefits or cost inputs, and the generalizability of the results.

Estimates are typically based on population means of sampled data; e.g., the cost in a specific health state, the QoL, the rate of clinical events over time, and relative effectiveness. Methodological uncertainty arises from differences in the choice of analytic—e.g., the perspective of the evaluation and discount rates. Structural uncertainty includes the models structure and interpretation, e.g., extrapolating costs over time.

4.5.5 HTA agencies lack of flexibility and different requirements from regulators

EU regulators have shown flexibility in assessing ATMPs focusing on the benefit-risk ratio more than the drug effect size, which constitutes the main focus of payers. Regulators have developed specific pathways to ensure faster approval of promising therapies including ATMPs (described in Chapter 3). The evidence considered sufficient by regulators is being scrutinized by HTA bodies and payers. They have expressed difficulties in assessing the value of cell and gene therapies due to uncertainties around long-term benefit and the demonstration of efficacy and safety on validated clinical outcomes.

So far, most HTA bodies seem reluctant to acknowledge the limitations in the applicability of their current analysis framework to cell and gene therapies. There is, from an HTA body's perspective, a clear fear that this practice may facilitate a multiplication of reference cases for a diverse class of products. HTA bodies tend to consider this new class on a case-by-case basis as they gain experience and may develop some specific methodology dedicated to ATMPs. Figure 4.1 presents the different perspectives of HTA agencies and regulators.

Regulators	HTA agencies/payers
Focus on assessment of drug benefit-risk	Focus on additional health benefits over the available therapies while accounting for budgetary constraints
Close attention to: • Quality • Efficacy and Safety (Internal validity) Evidence requirements differ between regulatory agencies (e.g. FDA vs EMA)	Close attention to: • Public health concerns and unmet medical needs • Relative effectiveness and safety (External validity) • Healthcare cost impact Various frameworks at national/regional/local levels
Setting up of new frameworks to enhance early access to innovative medicines, and acceptance of new study designs, e.g. • FDA breakthrough – single arm trials possibly sufficient • Adaptive pathways (AP) – pilot phase in Europe	Increased aversion to uncertainty and more stringent on evidence of efficacy and economic benefit • Want trials powered for economic & patient-relevant endpoints (e.g. OS, QoL) • Often will not extrapolate patient benefit outside the clinical trial population

Figure 4.1 Different evidence requirements between regulators and HTA agencies.

4.5.6 Conclusion on the valuation challenges

The majority of cell and gene therapies are developed to target severe or fatal diseases. Regulators and payers face substantial hurdles to find the right balance between authorizing early access and waiting for more clinical evidence, and delaying patients' access to innovation.

Cell and gene therapies are likely to face different challenges at the evidence generation level. For some, clinical evidence will necessarily come from small, single-arm, early phase clinical trials. In other cases, the identification of an appropriate comparator may be difficult, or randomization to a control group may be unethical. Reliance on evidence from single-arm trials or other observational evidence may, therefore, be required. The use of surrogate outcomes rather than clinical outcomes, e.g., progression-free survival, is a source of considerable uncertainty because they may not be able to translate to benefits for a clinical endpoint.

The implications of the availability of limited clinical evidence will largely depend on the HTA body, the level of unmet need in the target population, and the transformative potential of the cell and gene therapy. While uncertainty is unavoidable, there are ways to manage this uncertainty to provide suitable incentives for both developers and health insurance. These ways will be detailed in the next chapter.

4.6 Commercial challenges

In addition to the challenges at HTA and reimbursement level, cell and gene therapies face hurdles in commercialization. One of the main reasons may be that the majority of cell and gene therapies are developed by academics, biotechs, or small- and medium-sized enterprises with limited experience in product commercialization [26,27]. A study in the UK investigating barriers in commercialization of regenerative medicine showed that there are several barriers such as scaling up, lack of experienced people, and lack of business models [28]. The low level of commercialization of cell and gene therapies has been attributed to a variety of additional factors, such as the complexity in manufacturing processes and chain of distribution [29].

4.7 The consequences of market access delays

Market access delays led to delays in return on investment of the manufacturers that started to face economic difficulties. Indeed, 5 out of the 14 MA granted for ATMPs in the EU have been withdrawn so far (Chondrocelect®, Glybera®, Provenge®, MACI®, and Zalmoxis®) due to limited access to the market (Table 4.4). To illustrate the hurdle in commercializing new cell and gene therapies, a case study of an approved cell therapy, Provenge®, will be detailed.

Table 4.4 Reasons for EU marketing authorization withdrawal

ATMPs	Reasons for Withdrawal
Chondrocelect®	Approved in 2009, Chondrocelect® could not obtain reimbursement in the key EU countries. Therefore, the manufacturer, Tigenix, decided to withdraw the MA due to commercial reasons. Continuing to invest in an approved ATMP and pay to provide additional data is associated with uncertainties whether payers will approve it for reimbursement after evidence generation. Tigenix decided to concentrate its resources and capabilities on its allogeneic stem cell platforms, its upcoming Cx601 Phase III US trial, and its other clinical stage assets [30].
MACI®	Due to commercial reasons, MACI® marketing authorization holder closed the manufacturing site in Denmark in September 2014. Consequently, the manufacturing license was withdrawn. The marketing authorization was suspended on November 19, 2014, until a new manufacturing site is registered in the EU [31].
Glybera®	In April 2017, Glybera® MA holder—Uniqure—announced that the MA renewal for Glybera® would not be pursued. Therefore, the MA of this first gene therapy will expire in October 2017. This withdrawal was due to the limited market access of Glybera® and not related to any risk-benefit concerns. Uniqure decided to focus its financial resources on the development of another therapy [32].
Provenge®	The manufacturer of Provenge®, Dendreon, went bankrupt due to commercial reasons. Multiple factors contributed to the bankruptcy of Dendreon such as the market access barriers and the high price of Provenge® [33]. On May 6, 2015, the European Commission withdrew the marketing authorization for Provenge® after the request of Dendreon [34].
Zalmoxis®	The Company's decision takes into account the overall results of the interim analysis voluntarily carried out by the Company as part of the review of the product development plan, as well as the interactions with EMA in the latest months.The Company will direct its investments and resources, that were destined to Zalmoxis®, to activities that better contribute to its business objectives [35].

4.7.1 Case study: Provenge®

Provenge® was developed by Dendreon to treat patients with metastatic advanced prostate cancer. It was the first vaccine for cancer, first alternative to chemotherapy for patients with prostate cancer, and the first personalized medicine for cancer.

In May 2007, FDA in the US rejected Dendreon's application considering that Provenge® did not achieve the primary endpoints (progression-free survival) in the two Phase III trials.

In 2010, Provenge® was approved by the FDA for the treatment of asymptomatic or minimally symptomatic mCRPC based on a new clinical trial that showed 4 months' survival gain versus placebo (25.8 versus 21.7 months) and no improvement in progression-free survival.

In September 2013, Provenge® was granted an MA in the EU with a narrower indication: "patients with non-visceral metastases and who have not yet received chemotherapy." Provenge® was only reimbursed in Germany; it was not recommended for reimbursement in the UK.

Provenge® was priced at $31,000 per infusion in the US at the time of launch; the entire cost of treatment is $93,000 for three doses. In March 2011, CMS approved reimbursement of Provenge®, which simplified reimbursement in the whole labeled indication.

The main issue was the slow uptake of the drug; US oncologists and urologists said reimbursement issues have contributed to the slow uptake of the treatment, but the complexity of administering the drug is also a deterrent; in addition, Provenge® was considered expensive. Furthermore, Provenge® sales declined in the quarter immediately following the approval of Zytiga® in the pre-chemotherapy setting.

Dendreon's predictions from the sales of Provenge® were $4 billion per year. It achieved net revenue of $300 million from the sale of Provenge® in 2014, compared to $283.7 million in 2013 and $325.3 million in 2012. The company had $2.3 billion in losses.

As a consequence, Dendreon withdrew Provenge® MA in the EU, and Dendreon was acquired by Valeant in March 2015 [33]. Provenge® is still available in the US.

In June 2017, Valeant completed the sale of Dendreon to the Sanpower Group for $819.9 million in cash, as part of a realignment of its product portfolio strategy [36].

4.8 Conclusion

Until early 2019, the value assessment was a main challenge for gene and cell therapies. Clinical trial designs have raised concerns regarding the uncertainty of the short- and long-term efficacy and safety for most gene and cell therapies. Various approaches were used by different HTA bodies to minimize the potential risk of accepting costly gene therapies with uncertain outcomes. The successful HTA experience of the two CAR-T cell therapies and Luxturna® in 2019 showed that HTA agencies in most countries have started to consider solutions for dealing with uncertainty; they have recommended these therapies despite the limitations in clinical evidence at time of launch. A reassessment of real-world data and long-term evidence is usually required.

References

1. Carr DR, Bradshaw SE. Gene therapies: The challenge of super-high-cost treatments and how to pay for them. *Regenerative Medicine*. 2016;11(4):381–93.
2. Prasad V. Immunotherapy: Tisagenlecleucel—The first approved CAR-T-cell therapy: Implications for payers and policy makers. *Nature Reviews Clinical Oncology*. 2018;15(1):11–12.
3. Macaulay R. Advanced therapy medicinal products-transformational patient benefits but destined for commercial failure? *Value in Health*. 2017;20(9):A702.
4. Hettle R, Corbett M, et al. The assessment and appraisal of regenerative medicines and cell therapy products: An exploration of methods for review, economic evaluation and appraisal. *Health Technology Assessment (Winchester, England)*. 2017;21(7):1–204.
5. HAS. CHONDROCELECT (chondrocytes autologues). 2013 [accessed on March 10, 2019]. Available from: https://www.has-sante.fr/portail/jcms/c_1615035/en/chondrocelect-chondrocytes-autologues.
6. HAS. HOLOCLAR (amplified autologous limbal stem cells), stem cells for autologous transplant. 2017 [accessed on March 10, 2019]. Available from: https://www.has-sante.fr/portail/jcms/c_2661389/en/holoclar-amplified-autologous-limbal-stem-cells-stem-cells-for-autologous-transplant.
7. HAS. YESCARTA (axicabtagene ciloleucel), CAR T anti-CD19. 2019 [accessed on March 10, 2019]. Available from: https://www.has-sante.fr/portail/jcms/c_2888882/fr/yescarta-axicabtagene-ciloleucel-car-t-anti-cd19.
8. HAS. KYMRIAH (tisagenlecleucel), CAR T anti-CD19 (LAL). 2019 [accessed on March 10, 2019]. Available from: https://www.has-sante.fr/portail/jcms/c_2891689/fr/kymriah-tisagenlecleucel-car-t-anti-cd19-lal.
9. HAS. KYMRIAH (tisagenlecleucel), CAR T anti-CD19 (LDGCB). 2019 [accessed on March 10, 2019]. Available from: https://www.has-sante.fr/portail/jcms/c_2891692/fr/kymriah-tisagenlecleucel-car-t-anti-cd19-ldgcb.
10. G-BA. Zusatznutzen von Glybera® (Wirkstoff: Alipogentiparvovec) nicht quantifizierbar. 2015 [accessed on March 10, 2019]. Available from: https://www.g-ba.de/institution/presse/pressemitteilungen/578/.
11. G-BA. Nutzenbewertungsverfahren zum Wirkstoff Sipuleucel-T. 2015 [accessed on March 10, 2019]. Available from: https://www.g-ba.de/informationen/nutzenbewertung/143/.
12. G-BA. Nutzenbewertungsverfahren zum Wirkstoff Talimogen laherparepvec. 2016 [accessed on March 10, 2019]. Available from: https://www.g-ba.de/informationen/nutzenbewertung/243/.
13. Schuster SJ, Svoboda J et al. Chimeric antigen receptor T cells in refractory B-cell lymphomas. *New England Journal of Medicine*. 2017;377:2545–2554.

14. National Institute For Health And Care Excellence: Final appraisal determination Sipuleucel-T for treating asymptomatic or minimally symptomatic metastatic hormone-relapsed prostate cancer. 2015 [accessed on March 10, 2019]. Available from: https://www.nice.org.uk/guidance/ta332/documents/prostate-cancer-metastatic-hormone-relapsed-sipuleucelt-1st-line-id573-final-appraisal-determination-document2.
15. NICE. Melanoma (metastatic)—talimogene laherparepvec [ID508]. 2016 [accessed on March 10, 2019]. Available from: https://www.nice.org.uk/guidance/indevelopment/gid-tag509.
16. NICE. Holoclar. 2017 [accessed on March 10, 2019]. Available from: https://www.nice.org.uk/guidance/ta467/chapter/1-Recommendations.
17. NICE. Strimvelis for treating adenosine deaminase deficiency–severe combined immunodeficiency—Highly specialised technologies guidance [HST7]. 2018 [accessed on March 10, 2019]. Available from: https://www.nice.org.uk/guidance/hst7.
18. NICE. Tisagenlecleucel for treating relapsed or refractory B-cell acute lymphoblastic leukaemia in people aged up to 25 years-Technology appraisal guidance [TA554]. 2019 [accessed on March 10, 2019]. Available from: https://www.nice.org.uk/guidance/ta554/chapter/1-Recommendations.
19. NICE. NICE recommends another revolutionary CAR T-cell therapy for adults with lymphoma. 2019 [accessed on March 10, 2019]. Available from: https://www.nice.org.uk/news/article/nice-recommends-another-revolutionary-car-t-cell-therapy-for-adults-with-lymphoma.
20. NICE. Axicabtagene ciloleucel for treating diffuse large B-cell lymphoma and primary mediastinal large B-cell lymphoma after 2 or more systemic therapies—Technology appraisal guidance [TA559]. 2019 [accessed on March 10, 2019]. Available from: https://www.nice.org.uk/guidance/ta559.
21. ICER. Institute for Clinical and Economic Review. 2019 [accessed on March 10, 2019]. Available from: https://icer-review.org/.
22. Valuing a Cure: Draft Methods Adaptations ICER website. 2019 [accessed on September 9, 2019]. Available from: https://icer-review.org/material/valuing-a-cure-final-white-paper-and-methods-adaptations/.
23. CMS: Innovative treatments call for innovative payment models and arrangements, August 2017. Available from: https://www.cms.gov/newsroom/press-releases/cms-innovative-treatments-call-innovative-payment-models-and-arrangements.
24. ICER. Chimeric Antigen Receptor T-Cell Therapy for B-Cell Cancers: Effectiveness and Value. 2018 [accessed on March 10, 2019]. Available from: https://icer-review.org/wp-content/uploads/2017/07/ICER_CAR_T_Final_Evidence_Report_032318.pdf.
25. CMS. CMS proposes Coverage with Evidence Development for Chimeric Antigen Receptor (CAR) T-cell Therapy. 2019 [accessed on March 10, 2019]. Available from: https://www.cms.gov/newsroom/press-releases/cms-proposes-coverage-evidence-development-chimeric-antigen-receptor-car-t-cell-therapy.
26. Hanna E, Remuzat C, Auquier P, Toumi M. Advanced therapy medicinal products: Current and future perspectives. *Journal of Market Access & Health Policy*. 2016;4:31036.

27. ten Ham RMT, Hoekman J, Hövels AM, Broekmans AW, Leufkens HGM, Klungel OH. Challenges in advanced therapy medicinal product development: A survey among companies in Europe. *Molecular Therapy—Methods & Clinical Development*. 2018;11:121–30.
28. Plagnol AC, Rowley E, Martin P, Livesey F. Industry perceptions of barriers to commercialization of regenerative medicine products in the UK. *Regenerative Medicine*. 2009;4(4):549–59.
29. Yu TTL, Gupta P, Ronfard V, Vertès AA, Bayon Y. Recent progress in European advanced therapy medicinal products and beyond. *Frontiers in Bioengineering and Biotechnology*. 2018;6:130.
30. Tigenix. TiGenix reconfirms its strategic focus on its allogeneic stem cell platforms. 2016 [accessed on March 10, 2019]. Available from: https://www.marketwatch.com/press-release/tigenix-reconfirms-its-strategic-focus-on-its-allogeneic-stem-cell-platforms-2016-07-05-2184927.
31. EMA. MACI: Closure of EU manufacturing site. 2019 [accessed on March 10, 2019]. Available from: https://www.ema.europa.eu/en/medicines/human/referrals/maci.
32. Uniqure. uniQure Announces It Will Not Seek Marketing Authorization Renewal for Glybera in Europe 2017. Available from: https://www.globenewswire.com/news-release/2017/04/20/962549/0/en/uniQure-Announces-It-Will-Not-Seek-Marketing-Authorization-Renewal-for-Glybera-in-Europe.html.
33. Jaroslawski S, Toumi M. Sipuleucel-T (Provenge®)-autopsy of an innovative paradigm change in cancer treatment: Why a single-product biotech company failed to capitalize on its breakthrough invention. *BioDrugs: Clinical Immunotherapeutics, Biopharmaceuticals and Gene Therapy*. 2015;29(5):301–7.
34. EMA. Provenge: Withdrawal of the marketing authorisation in the European Union. 2015 [accessed on March 10, 2019]. Available from: http://www.ema.europa.eu/docs/en_GB/document_library/Public_statement/2015/05/WC500186950.pdf.
35. Molmed Press Release: MolMed informs on its decision to withdraw the Conditional Marketing Authorization of Zalmoxis, Milan, October 10, 2019. Available from: https://www.molmed.com/en/node/558.
36. Valeant Pharmaceuticals Completes Sale of Dendreon to Sanpower Group. 2017 [accessed March 10, 2019]. Available from: https://ir.bauschhealth.com/news-releases/2017/06-29-2017-120234317.

chapter five

How to mitigate uncertainties and HTA risk-averse attitude?

5.1 Conditional reimbursement

In the traditional reimbursement process, the HTA bodies assess the product clinical and economic value to issue recommendations on allowing or not allowing a new medicine to enter the health insurance system, therefore ensuring patients' access to this medicine or deny it. HTA bodies and payers have the responsibility of finding a balance between ensuring early patient access to transformative therapies with uncertain therapeutic value and delaying access while further data are collected.

The conditional reimbursement is one of the solutions proposed in the literature to mitigate cell and gene therapies' HTA challenges, by helping to deal with uncertainty at the time of launch. It aims to ensure the access of patients to new effective therapies, yet associated with high uncertainty; therefore, it aims to limit access delays while further evidence is collected [1,2]. This policy can make new promising cell and gene therapies available to patients at an early stage, and the final reimbursement decisions can be postponed until more robust evidence has become available [3].

Conditional reimbursement consists of an agreement between manufacturers and payers that clarifies the monitoring and controlling the uncertainty tools and defines the duration of this agreement. During this period, the manufacturer collects further data that limits the uncertainty, especially around long-term efficacy or safety, during reassessment.

According to the Lexchin study [4], conditional reimbursement or coverage with evidence development (CED) needs to be considered for:

- Expensive drugs with available data on intermediate endpoints,
- Drugs with the potential for widespread use but efficacy and/or safety is disputed,
- Drugs where RCTs' patient populations are small and are not representative of the target population.

The concept of conditional reimbursement or CED exists and varies across countries and reimbursement systems. It is called:

- "interim funding" in Australia,
- "conditionally funded field evaluation" in Ontario (Canada),
- "conditional reimbursement" in the Netherlands,
- "still in research" in France, and
- "monitored use" in Spain.
- There is also the Cancer Drugs Fund (CDF) for oncology drugs in the UK for CED reimbursement.

All these programs consist of ongoing data collection while the medical technology is already being funded by the healthcare system.

The availability of registries is of great importance for the purposes of monitoring conditional reimbursement agreements.

5.1.1 *Limitation*

In practice, gathering the additional evidence requires burdensome administrative procedures. In addition, once a medicine is used in practice, ending reimbursement may be less feasible than deciding not to reimburse in the first place—in particular, in case this medicine has proven to be effective, but it was not considered cost effective [2]. Therefore, there is a need to clearly pre-specify the duration, the required type evidence to be collected during the period of conditional coverage, and the conditions under which conditional reimbursement can be considered a feasible or an optimal strategy.

5.1.2 *The use of conditional reimbursement*

Some countries—for example, the US, UK, Netherlands, and Sweden—have already implemented some form of conditional reimbursement [5].

- **Example of Netherlands:** In the Netherlands, the concept of conditional reimbursement was implemented in 2006 for new expensive inpatient drugs. In 2013, it was extended to include a selected group of outpatient drugs that met the criteria for temporary reimbursement. The duration of the agreement is four years. During this period, real-world data are collected. After four years, an evaluation is carried out to inform the final reimbursement decision [5].
- **The Italian model:** The data collection and conditional reimbursement: AIFA principle is that an innovative drug should be reimbursed only if it is effective. The welfare systems cannot assume the responsibility for the failures of ATMPs with high costs. There is a need for identification of responders that may benefit the most from the ATMP (Figure 5.1).

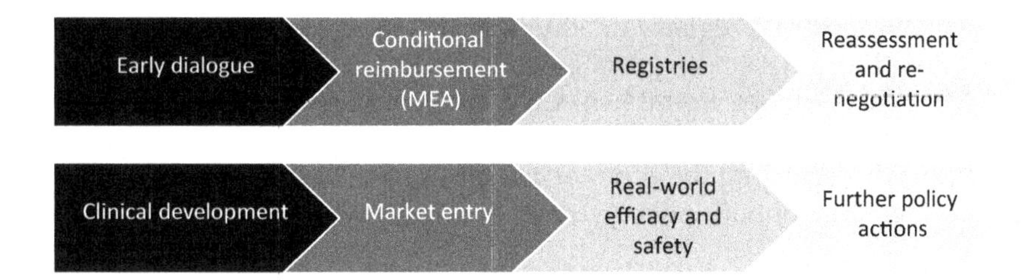

Figure 5.1 Italian model of conditional reimbursement. (Adapted from AIFA presentation Xoxi, E., Agenda item 4–Real world evidence data collection, Italian Experience on Registries, 2016.)

5.1.3 Conditional reimbursement for gene therapies

- **The CDF in England used for cancer drugs**
  The reformed Cancer Drugs Fund in England is a managed access fund, providing conditional coverage for cancer drugs where there is uncertainty in the economic case [6]. When considering a drug eligibility for the CDF, the NICE Technology Appraisal Committee will define and describe the area(s) of uncertainty.

  While the drug is in the CDF, data will be collected to resolve the uncertainty.

  CDF Commercial Agreements will be considered on a case-by-case basis so as to support the inclusion of cancer drugs in the CDF and facilitate patient access.

  When sufficient data are available to address the uncertainty, NICE will schedule the reappraisal of the drug. The NICE recommendation will either be a positive recommendation for routine commissioning or a negative routine commissioning recommendation [7].

  Two CAR-T (Kymriah® and Yescarta®) cells were reimbursed via the CDF so far.
- **Conditional reimbursement in France**
  In the HAS action plan for ATMPs published in January 2020, HAS considered that the top action will be the conditional reimbursement of ATMPs to mitigate uncertainty [8]. The ATMPs will be reimbursed while real world evidence long-term data is collected.

  The two reimbursed CAR-T cells on the market are reviewed annually unlike other pharmaceuticals.
- **The recommendation of CED for CAR-T cells in the US**
  On February 15, 2019, the Centers for Medicare & Medicaid Services (CMS) proposed Coverage with Evidence Development for CAR T-cell Therapy [9].

  The proposed National Coverage Determination would require Medicare to cover the therapy nationwide when it is offered in a

CMS-approved registry or clinical study, in which patients are monitored for at least two years post-treatment. Evidence from the registries and studies would help CMS identify the types of patients that benefit from CAR T-cell therapy, informing a future decision by the agency regarding the types of cases in which Medicare would cover the treatment with no registry or trial requirement.

5.2 *The use of registries to collect long-term, real-world data*

The registry studies help to continue the demonstration of the new therapy efficacy and safety in the real world. These studies allow identifying potential subpopulations and follow-on indications. In addition, by capturing the therapy outcomes, registries may support innovative payment models/contracting agreements. Registries are gaining increasing importance. In its action plan, the HAS highlighted the need for collecting real world evidence in national registries.

5.2.1 *EMA guidelines on establishing CAR-T patient registries*

In order to align patient registries established for the different CAR-Ts, the EMA's Patient Registries Initiative implemented guidelines for registries of patients who receive CAR-Ts. The first two CAR-Ts that will be included are Kymriah® and Yescarta® approved in August 2018. The long-term safety and efficacy of the therapies must be monitored through a patient registry. In its final report, EMA identified several key recommendations for using registry data to support risk-benefit evaluations of CAR-Ts. Patients must have follow-up visits at the treatment centers at three months, six months, one year, and then annually following the CAR-T. EMA recommended [10]:

- To harmonize a set of core, commonly defined data elements collected across registries
- To establish measures that ensure data are collected systematically with appropriate verification and quality assurance
- To ensure arrangements are in place to permit data sharing
- To improve communications between registry holders, regulators, and marketing authorization holders and applicants

5.3 *Improving the acceptability of single-arm with historical control trial design*

A randomized controlled clinical trial is the optimal design to evaluate the efficacy of a novel experimental therapy versus another therapy used in current practice. This trial design allows quantifying the added benefit of the new therapy while minimizing the biases and confounding factors.

However, in some cell and gene therapy cases, randomized trials are not feasible or unethical due to different reasons including the small target populations and personalized nature of therapies (such as autologous CAR-T cells). Single-arm trials are used in these cases.

Historical control based on real-world evidence may help to build the "missing arm" in the single-arm trials; it can potentially provide a reliable assessment of the safety and effectiveness of available therapies. External controls are often used to bridge the gap of providing comparative evidence using direct adjusted comparisons [11].

However, these studies typically do not use individual patient-level data, and there is a risk of systematic variation and biases. *Therefore, robust methods of gathering and assessing comparative evidence are needed to ensure acceptability of these methods by HTA bodies.*

To improve the acceptability of historical controls by HTA bodies, it is recommended:

- To use robust method to match populations, methods to control confounding factors, e.g., propensity scores;
- To assess the heterogeneity in the patient populations and its impact on the outcome;
- To identify all potential confounding factors;
- To understand patient pathways and disease management and make sure that it is well established and standardized;
- The primary endpoint needs to be objective and robust;
- To proactively assess the generalizability and transferability of the clinical data.

If the effect size of the new therapy is outstanding versus the historical cohort, there is a high chance that the historical cohort will be accepted by HTA bodies.

5.4 Recommendations of adaptation of economic evaluation

5.4.1 ICER initiative

On January 23rd, 2019, ICER launched a new initiative to develop and test alternative methods for the evaluation of potentially curative therapies and for translating the results of cost-effectiveness analyses into recommendations for value-based price benchmarks. ICER collaborated with health economists, stakeholders, and international HTA bodies, including NICE and the Canadian Agency for Drugs and Technologies in Health (CADTH). The aim of this project is to ensure that assessment methods are appropriate for the transformative nature of the potentially curative therapies [12] (Further details in Section 4.4.1).

5.4.2 Our suggestion for adaptation of economic modeling requirements

Some new methodological considerations need to be adopted taking into consideration the specificities of cell and gene therapies. Table 5.1 summarizes our recommendations for cell and gene therapies' economic models.

5.5 Early dialogue with regulators and payers

Evidence requirements for regulators are different from requirements for HTA agencies/payers. To bridge this gap, it is recommended to engage early in the development of the therapy with HTA agencies and regulators. Early dialogue allows pharmaceutical companies to gain critical insights from HTA bodies and/or regulators early in the development of a medicine, generally before initiation of a phase III clinical trial. It helps to ensure that their clinical development plan is appropriate from the HTA and regulatory perspective and to optimize evidence generation and address payers' needs. It also increases the chances of success at time of launch.

Different options of early advice exist currently:

- Parallel advice EMA-HTA: Scientific advice of EMA is sought in parallel with HTA bodies.
- Multi-HTA early advice: Early dialogues are done with the consortium of HTA bodies in the EUnetHTA.
- National HTA advice: It is an early dialogue with a single HTA body. Several countries put in place HTA early advice for clinical development plan of health technologies—e.g., UK, Germany, France, and Sweden.

A limited number of early dialogue activities are currently conducted; timelines are not aligned to the fast pace of development of cell and gene therapies. For example, ZIN (HTA body in Netherlands) provides early advice on 6–10 products per year through EUnetHTA.

5.5.1 Limitation

It is not a risk-neutral procedure: Early dialogue might increase the complexity of the development plan, and subsequent non-conformance with specific advice will have to be documented. The company needs to be well prepared to mitigate risk and to take advantage of the interactions with HTA agencies/EMA to maximize benefit.

5.5.2 Benefit-risk assessment for HTA early advice

Pharmaceutical companies need to balance the benefits and risks to seek early advice for their new therapy. Figure 5.2 illustrates the advantages and risks that need to be assessed by the company to decide on go/no-go for early dialogue.

Table 5.1 New methodological considerations for cell and gene therapies economic models

Valuation of benefits	Person-trade-off method (PTO) instead of Time-trade-off (TTO)	• Application of different weights to QALYs, or different valuations for specific groups
	Holistic valuation approaches	• Using more holistic valuation approaches instead of EQ-5D, with standardized methods to ensure some comparability between studies for different diseases
	Disease-specific valuation	• Considering potential disease specific valuation methods
	Caregiver utility	• Caregiver gain of utility should be considered in the economic model for cell and gene therapies.
	Curative value	• Value for cure should be integrated.
In the case of cell and gene therapies, there are potential important costs outside the health sector: social services, special education. Society perspective should be considered.	Future medical costs	• Future medical costs related and not directly related to the disease, incurred over life-years gained, should be considered.
	Future social costs	• Future social costs and education costs related and not related to the disease should also be included.
	Productivity costs	• Productivity costs should be taken into consideration. • Caregiver loss of productivity and time for involvement as well as cost should be included.
Time horizon	Different time horizons	• Present different time horizons: a clear rule on how to use multiple time horizons for decision-making ultimately is needed. • Time horizon should depend on the type of therapy and the disease and should not be set as one fit for all.

(Continued)

Table 5.1 (Continued) New methodological considerations for cell and gene therapies economic models

	Experts' engagement	• Neutral experts should be involved in the Delphi panel to provide the likelihood of long-term benefit over a range of time horizon to inform the modeling. • All models should undertake structured expert discussions to elicit probabilities associated with benefits and harms over different periods of time.
	Informed decision	• Make an informed decision about the appropriate time-horizon, rather than having a time horizon based on the beliefs of decision-makers who may not have the relevant expertise.
	Future outcomes	• Future outcomes should be weighted based on these elicited probabilities.
Discount rate	Lower discount rate	• Possibility of using different discount rates: • Discount rate for therapies with large benefit (40 QALYs) or for cures may be lower.
Payment models	Included in the economic model	• Payment models should be integrated in the cost-effectiveness modeling when considering drug cost. • The impact of a new payment model should be considered when quantifying the uncertainty.
Impact on disease prevalence	Disease prevalence will increase	• There is an increase of prevalence because patients will have longer life expectancy and have descendants who will transmit the disease unless pre-natal diagnostic is performed followed by abortion or fertilization in vitro on selected gametes.

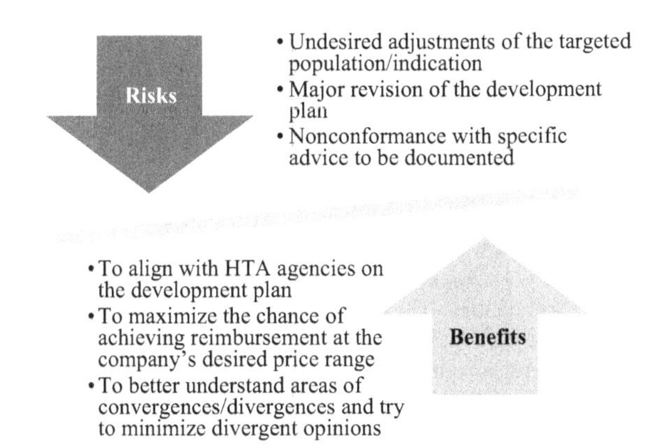

Figure 5.2 Benefit-risk assessment for HTA early advice.

The importance of early dialogue and collaboration with HTA bodies was confirmed by Luxturna®'s fast-track approval by NICE via a highly specialized technology procedure of 20 weeks instead of 38 weeks. Haseeb Ahmad, country president of Novartis UK, said: "Through effective collaboration with NICE and NHS England, it has been possible to secure rapid access to the first and only one-time gene therapy for patients living with this condition." In addition, Meindert Boysen, director of the Centre for Health Technology Evaluation at NICE, said: "The company's willingness to work with us early and constructively has allowed us to publish this guidance on a much faster timeline than normal which is good news for patients" [13].

5.6 Conclusion

An adaptation of the traditional HTA processes is required for ensuring patient access to cell and gene therapies without delays. HTA bodies are increasingly adopting conditional reimbursement to avoid delays in access while further data are collected. The real-world evidence data collected via registries are becoming of high importance.

Some HTA bodies have started to consider some revisions of the current economic evaluation framework to adapt it to innovative therapies. However, most HTA bodies remain conservative and consider their current analytic framework to be appropriate and not in need of updating to address these new therapies.

However, developers and HTA bodies started dialogue to find solutions to mitigate challenges at the HTA level.

HTA challenges are not the only challenges facing innovation; the high price of these therapies raises affordability issues that need to be addressed to ensure the sustainability of the healthcare system. HTA may be used as a cost-containment tool, as the HTA bodies assess the value of a new drug and its potential price.

References

1. Walker S, Sculpher M, Claxton K, Palmer S. Coverage with evidence development, only in research, risk sharing, or patient access scheme? A framework for coverage decisions. *Value in Health*. 2012;15(3):570–9.
2. van de Wetering EJ, van Exel J, Brouwer WBF. The challenge of conditional reimbursement: Stopping reimbursement can be more difficult than not starting in the first place! *Value in Health*. 2017;20(1):118–25.
3. Trueman P, Grainger DL, Downs KE. Coverage with evidence development: Applications and issues. *International Journal of Technology Assessment in Health Care*. 2010;26(1):79–85.
4. Lexchin J. Coverage with evidence development for pharmaceuticals: A policy in evolution? *International Journal of Health Services: Planning, Administration, Evaluation*. 2011;41(2):337–54.
5. Franken M, le Polain M, Cleemput I, Koopmanschap M. Similarities and differences between five European drug reimbursement systems. *International Journal of Technology Assessment in Health Care*. 2012;28(4):349–57.
6. Morrell L, Wordsworth S, Schuh A, Middleton MR, Rees S, Barker RW. Will the reformed cancer drugs fund address the most common types of uncertainty? An analysis of NICE cancer drug appraisals. *BMC Health Services Research*. 2018;18(1):393.
7. Cancer Drugs Fund NHS England website. Available from: https://www.england.nhs.uk/cancer/cdf/.
8. HAS, La HAS présente son plan d'action pour les médicaments innovants January 2020. Available from: https://www.has-sante.fr/jcms/p_3148262/fr/la-has-presente-son-plan-d-action-pour-les-medicaments-innovants.
9. CMS. CMS proposes Coverage with Evidence Development for Chimeric Antigen Receptor (CAR) T-cell Therapy CMS Website. 2019 [accessed on May 2, 2019]. Available from: https://www.cms.gov/newsroom/press-releases/cms-proposes-coverage-evidence-development-chimeric-antigen-receptor-car-t-cell-therapy.
10. EMA. Report on CAR T-cell therapy Registries Workshop February 9, 2018. Available from: https://www.ema.europa.eu/en/documents/report/report-car-t-cell-therapy-registries-workshop_en.pdf
11. Davies J, Martinec M, et al. Comparative effectiveness from a single-arm trial and real-world data: Alectinib versus ceritinib. *Journal of Comparative Effectiveness Research*. 2018;7(9):855–65.
12. ICER. ICER Launches International Collaborative to Develop New Methods to Guide Value-Based Pricing of Potential Cures. 2019. Available from: https://icer-review.org/announcements/icer-launches-international-collaborative-to-develop-new-methods-to-guide-value-based-pricing-of-potential-cures/.
13. NICE, NICE recommends novel gene therapy treatment for rare inherited eye disorder. Available from: https://www.nice.org.uk/news/article/nice-recommends-novel-gene-therapy-treatment-for-rare-inherited-eye-disorder.
14. Xoxi E. Agenda item 4–Real world evidence data collection: Italian Experience on Registries. 2016. Available from: https://ec.europa.eu/health/sites/health/files/files/committee/stamp/2016-03_stamp4/4_real_world_evidence_aifa_presentation.pdf.

chapter six

Cell and gene therapies funding

Challenges and solutions for patients' access

6.1 Current healthcare spending

Nowadays, the healthcare expenditure tends to increase; around 10% of the Gross Domestic Product (GDP) is spent annually on healthcare in the EU [3] and around 18% of GDP in the US (2016 data [4]). Based on OECD data [5], pharmaceuticals expenditure constitutes an important part of healthcare expenditure. After comparing the GDP, pharmaceutical expenditures and health expenditures excluding pharmaceuticals, based on OECD data, it was shown that both pharmaceutical and total health expenditures grew at a higher rate than the mean annual growth rate of GDP for the OECD countries between 2000 and 2017. Pharmaceuticals growth was higher than health expenditure between 2000 and 2006, and then the health expenditure growth surpassed that of pharmaceuticals (Figure 6.1).

The key factors of the rising expenditure on healthcare are population growth, population aging, disease prevalence or incidence, medical service utilization, and price.

Therefore, the different countries worldwide are struggling with the increase in health spending even before the arrival of a large number of gene and cell therapies to the markets.

In addition, most current financial systems are built on annual budgets. Paying for innovative therapies requires an allocation of resources from the healthcare/pharmaceutical budget [6]. Decision-makers have to scrutinize the incremental value of the new medicines, to make sure the budget is used to cover therapies that ensure optimal benefits for patients and society. If the healthcare budget is spent on drugs that have only marginal effect, less money will be available for transformative drugs.

The affordability will depend on the country GDP, in addition to the healthcare expenditure that helps define willingness to pay and budget availability.

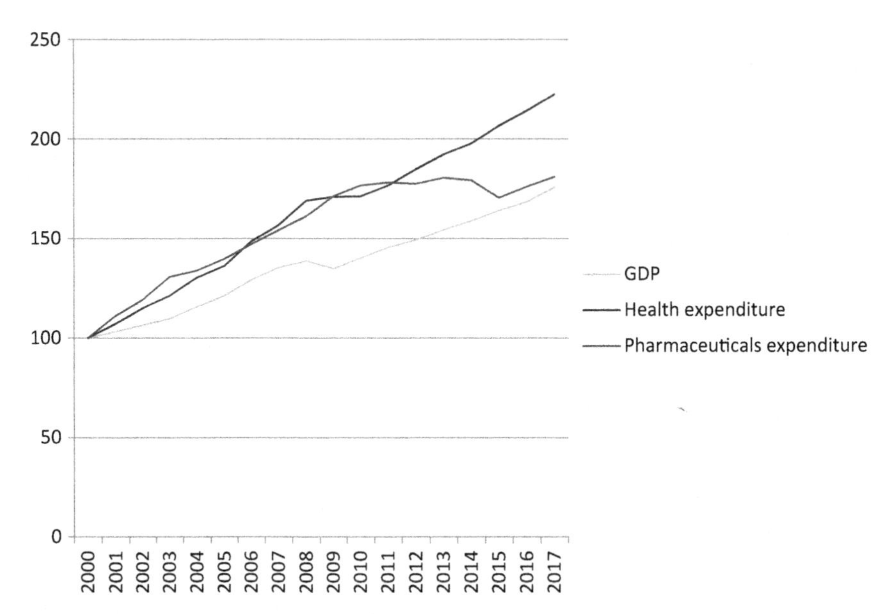

Figure 6.1 Trend growth in pharmaceutical expenditure, health expenditure, and GDP between 2000 and 2017.

It is generally anticipated that advanced therapies have the potential to dramatically change the way healthcare is provided and therefore to drive reallocation of significant resources. As a result of these expected changes, many stakeholders are concerned about affordability of advanced therapies and consequently, potential negative net effects on healthcare budgets.

Furthermore, there are different public health priorities; healthcare systems value health priorities differently and often reach conflicting decisions about treatments. Concerns around affordability of specific treatments may also influence health priorities and decisions regarding medicinal products.

6.2 *Current cost-containment policies*

Under the continued austerity, payers are trying to limit pharmaceutical spending by applying cost-control initiatives using a range of tools, including control of prices and/or volumes (e.g., price-volume agreements). Some policies control the level of spending for particular products (e.g., product-specific rebates) or for pharmaceuticals generally (e.g., claw-backs, patient cost sharing) [7].

Among the pricing policies used currently in the EU were free pricing, rate of return regulation, external reference pricing, cost plus pricing, clinical and cost-effectiveness pricing, value-based pricing and price revision (mandatory price cuts).

6.2.1 *External reference pricing*

External reference pricing (ERP) is defined by the WHO Collaborating Centre for Pharmaceutical Pricing and Reimbursement Policies as: "The practice of using the price(s) of a medicine in one or several countries in order to derive a benchmark or reference price for the purposes of setting or negotiating the price of the product in a given country" [8]. ERP is one of the most common cost-containment tools to reduce prices for in-patent pharmaceuticals in the EU member states [9]. Germany and UK are often first-launch countries; they allow free pricing for innovative drugs, they constitute together with France the three countries most commonly referenced. ERP is not conceptually considered an appropriate method to contain pharmaceuticals expenditure, but it could be an effective tool [9].

6.2.2 *Internal reference pricing*

This approach is mainly used to define the price or reimbursement price of generic drugs and, less commonly, therapeutic alternatives, at market entry. It consists of pricing drugs by reference to therapeutic comparators.

6.2.3 *Value-based pricing*

Value-based pricing (VBP) terminology means a price that reflects the value to the customer [10]; the price is linked to evidence-based assessments of value for patients and the society of a new therapy. It "sets prices based on a value assessment that takes into consideration a wider range of criteria than clinical cost–effectiveness, including the burden and severity of the disease and long-term benefits of the treatment" [7].

6.2.4 *Cost-based pricing*

Cost-based or cost-plus pricing is "a method for setting retail prices of medicines by taking into account production cost of a medicine together with allowances for promotional expenses, manufacturer's profit margins, and charges and profit margins in the supply chain" [11]. It is generally not recommended as an overall pharmaceutical pricing policy. It is used in some low- and middle-income countries—e.g., China, Vietnam, Bangladesh, Indonesia—and in high-income countries including Australia, Greece, and Spain, but it is usually confined to locally produced pharmaceuticals.

6.2.5 *Profit control as an indirect price control*

The UK uses indirect price control by limiting pharmaceutical companies' profits on UK territory. Manufacturers are free to set the price at market entry. Further increases are limited by the Pharmaceutical Price Regulation Scheme (PPRS). If a company's rate of profit exceeds the authorized level, it must reduce the general price level of its products in a way designed to pay back excessive returns to the NHS, but it remains free to decide on which products will see price reductions and to increase prices of other products.

6.2.6 *Discounts, rebates, expenditure caps, and price-volume agreements*

Discounts and rebates are price reductions and refunds linked to sales volume. They are being granted to public payers by pharmaceutical companies in 25 of the 31 European countries surveyed in a European survey (outpatient sector in 21 countries and inpatient sector in all 25 countries) [12].

An expenditure cap is a type of agreement between manufacturers and payers where the expenditures of the payers on a given drug are capped at a predefined amount. These agreements are usually put under risk-sharing agreements [13].

Pharmaceutical firms may negotiate based on the total value of sales, rather than on a per-unit price basis. Price reductions are obtained when volume increases through these agreements. The French pricing committee (Comité Economique des Produits de Santé: CEPS) sometimes enters into volume-price agreements for products with high sales potential; the "price reduction" takes the form of rebates, paid at the end of the year by the manufacturer with no consequences for the listed price.

6.3 *Proposed funding models for high-cost therapies*

The funding models proposed in the literature were classified in three categories: financial-based agreements, health outcomes-based agreements, and healthcoin (Figure 6.2). Funding models may be indication specific or not linked to the indication.

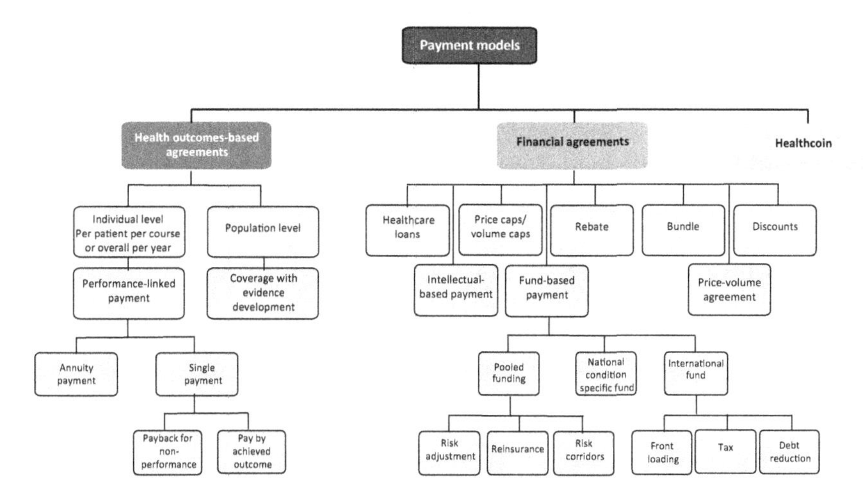

Figure 6.2 Funding models classification.

6.3.1 *Financial agreements*

These agreements between payers and manufacturers were based only on financial aspects and were independent of health outcomes of the novel therapy. The financial agreements identified were grouped as follows:

6.3.1.1 *Bundle payment, episode of care*

6.3.1.1.1 Definition An episode of care is a single payment for all clinically related services for one patient for a predefined discrete diagnostic condition. It is characterized by events defining the start and end dates [14,15]. A bundle payment is an integrated single payment that covers all healthcare services related to a specific treatment or procedure [14–21] instead of paying for every service provided. Episode of care and bundle payments can yield savings in three ways: (1) price negotiation so the total cost will be less than fee-for-service; (2) agreement with providers that any savings that arise will be shared with them; (3) savings because no additional payments will be made to treating complications of care.

This is also the principle of an integrated health system, where healthcare providers create a joint organization to deliver comprehensive care for patients with a given condition. In the US, integrated systems were promoted by Obamacare (the Patient Protection and Affordable Care Act) and are called Affordable Care Organizations (ACO).

For cancer care, a new model of ACO—the Oncology Care Model (OCM)—was developed by the Centers for Medicare & Medicaid Services (CMS). It is an episode-based model; the total cost includes chemotherapy treatment and all medical services during the following six months.

In addition to the fee-for-service payment for each episode of oncology care, the model includes two further payments [22]:

- **Funding for enhanced care management services:** This is a per beneficiary per month (PBPM) fee for each episode of chemotherapy.
- **Performance-based payment:** It is a semi-annual lump sum. It depends on satisfactory quality metrics and spending per chemotherapy episode falling below a predefined target.

6.3.1.1.2 Applicability for advanced therapies funding These models aimed to transfer the drug budget responsibility to healthcare providers with an opportunity for payers to avoid negotiating prices. Prices are discussed between manufacturers and healthcare providers. The payer shifts the financial risk to healthcare providers and creates incentives for budget control by providers. Performance criteria may be part of the quality metrics used by payers to create incentives for outcomes optimization. One of the limitations of this type of payment model is that when novel, innovative, potentially expensive therapies become available, there is a need for temporary funding on the top of the predefined payment. Usually such payments are covered by a central budget envelope. This may be an incentive to use expensive products funded by a side budget envelope. It will not be attractive or feasible in countries where changes of insurers or treating physicians are frequent, such as in the US. Under the provision that patients remain covered by the same insurer in the long run, such payment models may prove to be useful for cell and gene therapies funding.

6.3.1.2 Rebates and discounts

6.3.1.2.1 Definition Rebates are payments refunded by the manufacturer to the payer after the transaction has occurred [23,24]. This commercial agreement, usually confidential, is becoming increasingly popular in several countries. It may be driven by the incremental cost-effectiveness ratio (ICER) or result simply from a negotiation with no objective economic evidence to match the affordability or willingness to pay subjectively defined by the payer.

Discounts are price reductions granted to payers under specific conditions [25–27]. It is a simple discount on the unit price of a drug. They are usually confidential, and do not affect the list price of the drug. Discounts can vary widely, but most commonly the discount is over 20% of the list price [28]. Confidential discounts are now common in Europe.

6.3.1.2.2 Applicability for advanced therapies funding Discounts and rebates aim to reduce the bill on the payers and thus adjust the net price to payers' affordability while maintaining the facial/listed price. These agreements will temporarily be helpful in rich countries, but it is unlikely to help low- and middle-income countries to access such products that are by far unaffordable for them.

6.3.1.3 Price caps/volume caps per patients or per target population

6.3.1.3.1 Definition Price caps and volume caps are methods used to control and limit pharmaceuticals prices and manufacturers' revenues [23,24,29].

- At the patient level, they aim to cap the yearly price or the number of yearly treatment courses reimbursed. If additional courses are needed, these have to be provided by the manufacturer free of charge.
- At the population level, these strategies aim to cap the yearly expenditure/volume the manufacturer will be allowed to sell. Beyond the cap, the manufacturer may have to reimburse the full retail price, the full ex-factory price, or a proportion of the price, depending on the details of the agreement.

6.3.1.3.2 Applicability for cell and gene therapies funding It is unlikely to be the right solution for cell and gene therapies, as large diffusion to all eligible patients is the only ethically acceptable target.

6.3.1.4 Price-volume agreements

6.3.1.4.1 Definition These are agreements where drug prices are reduced based on sales volume. For example, after selling 10,000 vials, the price is reduced by 20% for the next vials and so on. Alternatively, depending on the total sales volume, the price will be discounted for all vials sold, according to a predefined scheme [24,30–32].

6.3.1.4.2 Applicability for advanced therapies funding They aim to provide a price volume indexation. Although not paying the same price for a larger or small volume of goods does make sense, it is unlikely to help making unaffordable prices affordable. Therefore, it may only marginally help to ensure access to cell and gene therapies. It will create inequity against small countries.

6.3.1.5 Healthcare loans/credits

Two frameworks are possible; credits [33–36] can be allocated to payers or patients.

- In the first framework, patients borrow money from a specific entity to pay their copayments. The loan is then amortized over a predefined period. It is the same concept as credit cards or student loans. The specific entity in this case may be from a pool of investors.
- In the second framework, private and public payers assume the debt.

The frameworks allow overcoming the financial liquidity limitation and increasing affordability. However, credits do not incentivize companies to lower the high cost of the drugs. Furthermore, payers' credits will increase the demand for the new drug, which will increase the burden on payers.

6.3.1.6 Fund-based payment divided in three subcategories: pooled funding, national silo funds, and special international fundraising

6.3.1.6.1 Reinsurance risk pool Pooled funding through the reinsurance risk pool of multiple payers [18,34,37–40] is a way to secure high costs of drug treatment for an individual outlier patient. Such insurance may apply to public-private partnerships that aim to recover the costs of goods within a given time period through the collaboration between public and private payers [18,38,40]. Zettler and Fuse Brown [34] proposed the 3Rs method: (1) risk adjustment is a fund collected from all payers that aims to compensate payers that incur unusually high costs; (2) reinsurance is an insurance policy that insurers buy to protect against excess financial risk; and (3) risk corridors method is when the U.S. Department of Health & Human Services (HHS) collects funds from plans with lower-than-expected claims and makes payments to plans with higher than expected [41–43].

Applicability for advanced therapies funding

Pooled funding and the 3Rs are interesting funding routes to ensure fair cost share between payers. They may address payers' hesitancy to invest when the return on investment is built over a long period. Therefore, it may help adopting advanced therapies in a fragmented volatile insurance market like the US. However, pooled funding may reduce the incentive of manufacturers to reduce advanced therapies' targeted prices. In addition, attendees of the ICER Policy Summit meeting stated that insurers are already trying to exclude gene therapies from reinsurance policies, although a limited number of gene therapies are currently marketed [44].

6.3.1.6.2 National silo funds National funds for specific conditions or diseases: Examples are the Cancer Drugs Fund (CDF) in the United Kingdom that pays for new cancer drugs rejected by NICE [41,42] and the Australian Complex Authority Required Highly Specialised Drugs Program that funds and delivers specialty drugs [43].

6.3.1.6.3 Special international fundraising Three international financing methods addressing noncommunicable diseases were identified [38]:

1. International taxes on specific transactions will be used to finance some drugs procurements and supply (e.g., airline tax levy).
2. Front-loading of aids from donors to invest in health programs or drugs (e.g., funds received by the International Finance Facility for Immunization)—these methods can ensure sustained and predictable annual funding.
3. Debt reduction where the creditor country writes off debt for low- and middle-income countries with the commitment of the latter to invest in a domestic therapy or prevention program—this method is a one-time transaction, unlikely to provide stable funding.

Applicability for advanced therapies funding

The levies, front-loading, and the debt reduction are alternative solutions that may be considered for low- or middle-income countries or countries affected by unmanageable debts that constraint dramatically their ability to invest in health. These methods are not specific to cell and gene therapies but may contribute to enhance ability to access some specific advanced therapies through international solidarity especially if associated with differential pricing. This is typically what is happening in Europe for all fields but health. It is hard to understand that such regulation is in place in the EU for compensating agriculture products but is considered impossible for health services or innovative expensive products. Sustainable funding is critical; an initial funding needs to be associated with follow-up funding.

6.3.1.7 *Intellectual-based payment*

This approach includes prizes for patents—public payers buy the therapy from the manufacturer and gain control over its production and distribution [37]. The approach may also involve out-licensing production and distribution to payers, with the manufacturer maintaining its intellectual property (IP) rights. Furthermore, prolonging patent rights—as with orphan drugs—rewards innovation. In Europe, orphan drugs benefit from a market exclusivity of 10 years after the marketing authorization, in addition to the data exclusivity.

Applicability for advanced therapies funding

The IP route seems inappropriate, as the regulator will bear the production and distribution burden. In addition, if the government becomes the operator, the negotiation with the pricing committee representing the government would be unhealthy. It may be more appropriate for the government to better protect IP emerging from academic and public research centers, as more than 50% of drugs are built based on their discoveries. This funding process will unlikely be adopted by many liberal countries; moreover, it may not enhance access to advanced therapies.

The detailed definition with the advantages and limitations of each method are presented in Table 6.1.

6.3.2 Health outcomes-based agreements

This class encompasses agreements between manufacturers and payers based on drug performance. It is divided in two groups: individual-level performance-based payment and population-level conditional reimbursement often called coverage with evidence development (CED).

6.3.2.1 Payment-for-performance

"Pay-for-performance" (P4P) is a solution to pay for innovative high-cost therapies proposed in 15 studies [13,24,27,31,46–57]. Towse's definition of performance-based agreement [13] is an agreement between payer and manufacturer where "the price level and/or revenue received is related to the future performance of the product in either a research or a real-world environment." P4P encompasses several types of agreements: payback for nonperformance, payment for management of events that the drug failed to prevent, payment for managing undesirable effects, payment for achieved outcome, etc. [53]. In the latter agreement type, each treatment is considered either a success or a failure, based on a predefined outcome measure (e.g., disease progression) and a predefined timing of outcome assessment (e.g., six months).

Rebate risk sharing proposed by Kleinke and McGee [39] consists of rebates to patients with large cost sharing after the completion of milestones or course treatment. This method can improve patients' adherence as the adherence to the treatment will be rewarded by the rebate.

A limit pricing approach based on health outcomes similar to the NICE threshold was proposed by Fuller and Goldfield [51]. It includes in the calculation the payment for outcomes such as avoided hospitalizations and outcomes as decreased mortality not easily translated in dollars. Performance milestones not met would convert into price reductions.

Applicability for advanced therapies funding

P4P agreements ensure market access for innovative promising therapies, demonstrate value, allow sharing the risk between payers and manufacturers, and limit total budget impact. However, they require

Table 6.1 Identified funding models based on financial advanced therapies agreements: definition, advantages, and limitations

Method	Advantages	Limitations	References
Bundle	• Transfer the financial risks and rewards of patient care to healthcare providers • Allow better predictability of budget spending • Can yield savings for payers	Healthcare provider may limit the adoption of innovative therapies to maximize its margin.	Korda and Eldridge [16] Robinson [17] Jain and Shrank [18] Licking and Garfield [19] Greenapple [20]
Episode of care	• Improved quality of care and decreased economic burden • Allow better predictability of budget spending	Paying for discrete episode might not control the total number of episodes and could encourage more episodes. Feasibility concerns and implementation challenges	Barinaga et al. [14] Hussey et al. [15]
Rebates	• Confidential • Ensure savings	May distort external reference pricing	Gavious et al. [23] Jarosławski and Toumi [24]
Discounts	• The most relevant solution for payers • May be simple, depending on pricing	May be difficult to keep confidential	Anastasaki et al. [25] Aggarwal et al. [27]
Credits (Payers level)	Amortizing the cost of treatment by purchasing drugs without paying the entire price up front	Pharmaceutical companies may want to gain immediate revenue to boost their return on investment. Will eat into future revenue and reduce future affordability to payers	Philipson [35] Gottlieb [36]

(Continued)

Table 6.1 (Continued) Identified funding models based on financial advanced therapies agreements: definition, advantages, and limitations

Method	Advantages	Limitations	References
Healthcare loans (HCL) for patients (Hybrid models)	Increase drug affordability	Publicly available data on student loans and other consumer financing might not fully capture the risks of HCLs. Additional mechanisms to improve feasibility of third-party borrowing by patients' family members, for example Will eat into future revenue and reduce future affordability to payers	Montazerhodjat et al. [33] Philipson [35]
Price caps (mild regulation)	• Increase in consumer surplus and in the number of patients with only a marginal effect on the revenues of the company • Patient welfare improvement • Do not stifle the economic incentive for drug innovation.	Direct negotiation may bring a significantly lower price for payers.	Levy and Nir [29]
Price volume	Promoting sustainability	Complex to break down into various subtypes of costs	Messori [30] Barry and Tilson [31] Toumi et al. [32]

(Continued)

Table 6.1 (Continued) Identified funding models based on financial advanced therapies agreements: definition, advantages, and limitations

Method	Advantages	Limitations	References
Fund-based payment	Attractive, as funding is secured for conditions that are covered	National healthcare providers and insurers are unlikely to risk such a high level of investment for unproven drugs.	Carr and Bradshaw [37]
Intellectual-based payment	• Reward and incentivize innovation • Manufacturer no longer needs to seek high prices for treatment.	• Do not reduce the uncertainty • Neither attractive to manufacturers nor to payers as interests become reversed (payers become responsible for production, among others), while the risk of monopolization can have impact on innovation over time	Carr and Bradshaw [37] Jain and Shrank [18]
Pooled funding	Spread the cost over everyone in the insurance pool rather than imposing an unreasonable financial burden on the patient	Lack of integration of monitoring structures with other donor initiatives	Jain and Shrank [18] Meghani and Basu [38] Beauliere et al. [40] Kleinke and McGee [39]
Levies	Provide sustained funding	Airline ticket levy Would affect the volume of airline travel dissipated	Meghani and Basu [38]
Front-loading	Predictable annual funding	3.5% interest rate and associated commission fees linked with the bond sales	

(*Continued*)

Table 6.1 **(*Continued*)** IIdentified funding models based on financial advanced therapies agreements: definition, advantages, and limitations

Method	Advantages	Limitations	References
Debt reduction	Ensure fund	Limited impact in low- and middle-income countries with low external debt • No recurrent long-term, stable funding because often one-time transactions	
3Rs	Protect insurers against adverse selection and consumers against destabilization of the insurance market and discriminatory health insurance practices **Risk corridors** limit both downside risk of losses and excess profits for health plans.	Political and legal Challenges	Zettler and Fuse Brown [34] Proach et al. [45]
Cancer drug fund	Pay for new cancer drugs rejected by NICE	Onco-exceptionalism, inefficient, no discounts	Jack [41] Mayor [42]
Australian complex Authority Required Highly Specialized Drugs Program	Balance the benefits, risks, and costs	—	Lu [43]

logistics and bureaucracy that are associated with additional costs and increase manufacturer and healthcare provider burden. In addition, there is no guarantee that the product will retain its market position after reassessment; reimbursement discontinuation or price-cut are also possible options. In some circumstances where payers cannot clearly identify the source of uncertainties, an individual P4P may be the solution. Payers are increasingly experimenting with P4P schemes [48]. P4P agreements may have a place in advanced therapies funding, but their role will be addressing payers' uncertainties rather than budget constraints' challenges.

6.3.2.2 *Annuity payment*

"Annuity-style payment" [37], "annuity with risk sharing" [58], "technology leasing reimbursement scheme" [59], and "high-cost drugs mortgages" [39] are terms used in several articles to describe the agreement between manufacturers and payers that aims to replace the high up-front cost with a stream of payments triggered, at the patient level, by the achievement of clinical milestones. However, an important challenge lies in defining the clinical milestones and endpoints, which may be critical.

Applicability for cell and gene therapies funding

The annuity payment may be considered as an attractive payment for cell and gene therapies because it spreads out the up-front payment and links the payment to predefined health outcomes. This model seems the ideal solution as it addresses at the same time the high-budget impact and the performance uncertainty. However, it should be seen as a loan, and the delayed cost for payers is slowly cannibalizing the future healthcare budget.

Annuity payment and payers credits may not necessarily be an actual good solution. In reality, they will ultimately book the future payers' budget and will challenge the sustainability of the health insurance unless future revenues are expected to increase significantly to cover the credit engaged for expensive products access or when the additional budget impact is considered as affordable by payers.

6.3.2.3 *Coverage with evidence development*

Coverage with evidence development (CED) is also a method suggested in four papers [13,32,54,60]. It is a conditional reimbursement linked to the collect of post-launch real-world data. Once that data is available, a price renegotiation occurs if the product does not meet the expectations. However, in most cases, there are no prior agreements on the interpretation of evidence and price revisions based on the evidence generated by manufacturers. A CED scheme may be complemented with an escrow agreement. The escrow agreement places the sales revenue in an independent bank account; at the end of the study, the collected money is released to the company if the results are positive and to the health insurance if the results are negative. The CED may be the right tool to address uncertainty but not the affordability. In addition, it may fail to capture the needed data to reduce the decision-making uncertainty—thus, the agreement may be terminated [53].

6.3.3 Healthcoin

Basu et al. [61], suggested a new tradable currency—"Healthcoin"—as a financing mechanism for breakthrough therapies. It converts the incremental outcomes produced by curative treatments to a common numeraire, such as life-years equivalents. It can be traded to dollars in the marketplace. Medicare would pay the private payer for a beneficiary who is transitioning to Medicare at the age of 65 years, if the private payer had previously invested in a cure for diabetes, for example. Healthcoin incentivizes private payers to invest in breakthrough treatments, especially in curative therapies that are in demand for the non-elderly. The model limitation is the assumption that the cure is permanent and applies equally to all ages.

Healthcoin could be applied between private payers and Medicare or also between private payers when a patient switches from one plan to another. This will expand the possibility to invest in long-term benefit despite patients possibly switching after receiving the intervention.

6.4 Comparison of the proposed funding models features

Table 6.2 shows the funding model features: feasibility, acceptability, burden, financial attractiveness, appeal to payers, and appeal to manufacturer. Most of the models were considered feasible. The least feasible and acceptable ones were credits for patients, cost-plus pricing, intellectual-based payment, pooled funding, international funds, and healthcoin. Almost all of the funding models are associated with additional burden, except for rebates, discounts, price caps, and price-volume agreements. All models except intellectual-based funding could be considered appealing for payers, and the models most appealing for manufacturers were credits/HCL, national silo funds, international funds, and CED.

Figure 6.3 shows that annuity payment, P4P, discounts/rebates and national condition-specific funds are the top four funding models based on the feasibility and financial interest.

Figure 6.4 shows that national silo funds are the most appealing for both manufacturers and payers.

Table 6.2 Funding models features

Payment Models	Feasibility	Acceptability	Burden	Appeal (payers)	Appeal (manufacturer)
Financial-based funding					
Bundle payment/Episode of care	+++	+++	+	+++	++
Rebates/Discounts	+++	+++	x	+++	++
Price caps/volume caps	+++	+++	x	+++	+
Price-volume agreements	+++	+++	x	+++	++
Credits for patients	+	x	++	+	+++
Healthcare loans for payers	++	++	++	+	+++
National silo fund (e.g., CDF)	+++	++	+	+++	+++
Intellectual-based payment	+	x	+++	+	x
Pooled funding	+	+	++	+++	+++
3Rs	++	+	++	+++	+++
International funds	++	+	++	++	+++
Health outcomes-based funding					
Coverage with evidence development	+++	+++	+++	+++	+++
Pay for performance	+++	+++	+++	+++	++
Annuity payment based on performance	+++	+++	+++	+++	++
Healthcoin	+	x	+++	+	+

+: Low importance, ++: Important, +++: Very important, x: No

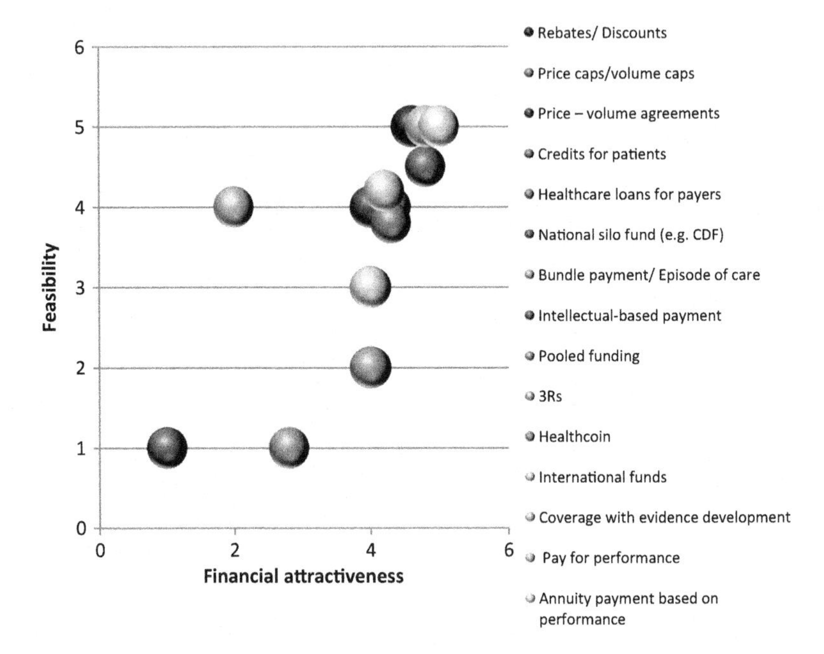

Figure 6.3 Matrix for the feasibility and financial interest of each funding model.

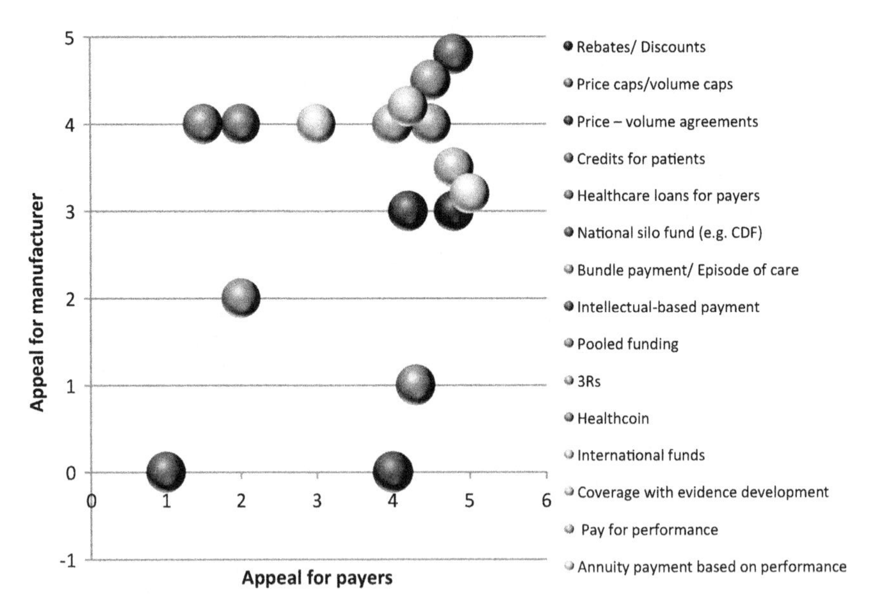

Figure 6.4 Matrix for the appeal for payers and manufacturer of each funding model.

6.5 Examples of adoption of new payment models

Some countries such as Italy have extensive experience in using MEA such as payment by results schemes for novel therapies. However, other countries such as Germany have rarely used such agreements. Recently, some health insurers in Germany have shown some flexibility in adopting new payment models.

6.5.1 Suggestion of a new pricing scheme in Germany

A health insurance in Germany, suggested a new price system called "dynamic evidence pricing" for gene therapies to prevent the "massive cost explosion": the price will be renegotiated two years after first negotiation, based on further evidence collected. After this renegotiation, the price may increase if the benefit is proven in real life and may decrease if the gene therapy fails to perform as claimed in the first submission.

6.5.2 Germany is piloting a pay for performance scheme

German health insurance started to pilot pay for performance agreements with developers of gene therapies. These agreements are confidentially agreed between developers and health insurance, they are based on achieved clinical outcomes. Clinical outcomes and duration are defined during the agreement.

6.5.3 Pilot test for annuity payment in the US

Pharmaceutical companies started discussing a pilot for annuity payment with commercial payers in the US for gene therapies. The prices are expected to be very expensive from $2 million to $5 million. Annuity payment will spread out insurance payments for high-cost treatments over a series of years. The model can also be linked to performance outcomes. The endpoints will be defined with the payers. The first payment would be up front. The participation of multiple payers would address the challenges with the risk of switching insurers in the US.

6.6 Review of the successful examples of CAR-T cells

The two CAR-T cells approved in 2018, Kymriah® and Yescarta®, are good examples of successful gene therapies that achieved reimbursement in the US and several EU countries. These products launches were highly prepared by manufacturers and anticipated by patients, medical professionals, and healthcare system stakeholders. They benefited from early access in some countries such as France through authorization for temporary use (ATU); in addition, outcomes-based agreements were achieved in several countries between manufacturers and payers.

6.6.1 France

Both Kymriah® and Yescarta® benefited from ATU in France. This scheme allows early access to the market for promising products before marketing authorization. It helps to collect real-world evidence on safety and efficacy in clinical practice. During this period, both CAR-T cells prices are set freely by manufacturers, while a maximum price is set and published by the pricing committee (CEPS).

After HTA recommendations (details in Chapter 4), TC requested to review annually these CAR-T cells with further long-term clinical data in addition to data from the French registry (through the Lymphoma Academic Research Organization [LYSARC] data platform) to address the uncertainties around the long-term efficacy, safety, and complexity of the treatment process and to assess the duration of production.

Yescarta® and Kymriah® are funded through the supplementary list of costly medicines (liste en sus), funded on top of DRG with a price of 327,000€ [62] and 297,666€, respectively.

6.6.2 England

They are both funded via CDF, which allows patient access to these promising new cancer treatments while more real-world evidence is gathered on effectiveness [63].

Additionally, real-world evidence will be collected for both CAR-T cell therapies through the Systemic Anti-Cancer Therapy (SACT) dataset and Bluteq, as well as through other Public Health England (PHE) datasets [64,65]. This combination of data sources will allow NICE and NHS England to more firmly establish the long-term efficacy (through the trial follow-up data) as well as the (shorter-term) real-world effectiveness in the English setting.

6.7 Future trends: which funding model could be adopted for advanced therapies?

The concept of "curative" therapies is getting close to reality, with the large number of advanced therapies in development. Questions on the adoption of these therapies, and on the means through which healthcare payers can finance them, are being raised increasingly. A specific financing issue differentiates potentially curative therapies such as advanced therapies from other pharmaceuticals; it is the high up-front costs with downstream long-term outcomes. These high up-front costs will likely threaten the sustainability of the healthcare system, and balancing innovation and affordability may become a challenge for payers.

6.7.1 *A potentially sustainable short-term solution: advanced therapies fund*

There are no perfect models. A likely sustainable model for advanced therapies may be an "advanced therapies-specific fund" independent from the traditional existing reimbursement path and independently funded. It is authors' suggestion as potential solution which consists of a special fund associated with other funding models to ensure its sustainability. Lessons can be learned from the CDF experience in the UK to avoid repeating the same errors, and provide rapid sustained access to advanced therapies without reaching an unbearable stage [66]. The experience of special funds was already adopted in several countries that have implemented special funds for orphan drugs. In Italy, a "5% AIFA fund" is collected from pharmaceutical companies, and half of the fund is devoted to providing access to medicines for rare diseases before marketing authorization [67]. In Scotland, "new medicines fund" (NMF) is implemented with £80M budget to ensure patients' access to end-of-life or orphan medicines while health boards are protected from the budget impact of funding these medicines [68].

In this model, funds sources would be tax based. Eligibility criteria should be clearly defined to restrict the entry for highly innovative, effective advanced therapies with high up-front cost and downstream delayed dramatic outcomes. A "cell and gene therapies value assessment framework" may be needed. The ICER threshold may be revised down while the discount rate may be revised up—for example, to adjust prices of cell and gene therapies to affordability (size of the fund). Ultimately, the investment in advanced therapies will also need to be contained within a predefined limited budget. Therefore, all cost-containment measures could be applied. The various discounts/rebates/price volume agreement/cost sharing/annuity payments may apply to render those products more affordable, but this may certainly not be the cornerstone of funding cell and gene therapies.

To avoid the uncertainty about the added value of funded therapies such as with CDF—which was considered a waste of money by some authors [69,70]—CED with an escrow agreement should be the rule, as most of those highly promising therapy will reach the market with immature data. Such a scheme would allow maintaining a major pillar of innovation: the value-based pricing. A robust and effective horizon scanning will be a critical tool to allow a strong forecasting of requested resources to fund advanced therapies or to secure prices of advanced therapies are aligned with a budget. Finally, the government will need to define a maximal proportion of GDP allocated to this fund.

It is critical to keep in mind that by channeling an excessive budget to healthcare interventions, resources are displaced from health determinants (e.g., clean fresh water, education, social services, pollution control, etc.), which may contribute to citizen health worsening. This may be a health disinvestment rather than a health investment. The choice of the displaced resource allocation will be as important as the new funded intervention.

References

1. CMS: Innovative treatments call for innovative payment models and arrangements [press release]. Available from: https://www.cms.gov/newsroom/press-releases/cms-innovative-treatments-call-innovative-payment-models-and-arrangements.
2. Miller S. Gene therapy holds great promise, But big price. 2017. Available from: http://lab.express-scripts.com/lab/insights/drug-options/gene-therapy-holds-great-promise-but-big-price?hootPostID=6df8ced02008c379691d79549b166eaf&utm_source=STAT+Newsletters&utm_campaign=e308efdff8-EMAIL_CAMPAIGN_2017_09_20&utm_medium=email&utm_term=0_8cab1d7961-e308efdff8-149537785.
3. OECD. Health spending. Available from: https://data.oecd.org/healthres/health-spending.htm#indicator-chart.
4. Probasco J. Why do healthcare costs keep rising? 2018 [accessed on April 29, 2019]. Available from: https://www.investopedia.com/insurance/why-do-healthcare-costs-keep-rising/.
5. OECD. Pharmaceutical spending. Available from: https://data.oecd.org/healthres/pharmaceutical-spending.htm.
6. Hollis A. Sustainable financing of innovative therapies: A review of approaches. *Pharmaco Economics*. 2016;34(10):971–80.
7. WHO. Access to new medicines in Europe: Technical review of policy initiatives and opportunities for collaboration and research. 2015.
8. WHO. Collaborating centre for pharmaceutical pricing and reimbursement policies. [May 14, 2017]. Available from: http://whocc.goeg.at/Glossary/PreferredTerms/External%20price%20referencing.
9. Rémuzat C, Urbinati D, Mzoughi O, El Hammi E, Belgaied W, Toumi M. Overview of external reference pricing systems in Europe. *Journal of Market Access & Health Policy*. 2015;3. doi:10.3402/jmahp.v3.27675.
10. Liozua S, Hinterhuberb A, Bolanda B, Perellia S. The conceptualization of value-based pricing in industrial firms. *Journal of Revenue and Pricing Management*. 2011;11(1):12–34.

11. Cameron A, Hill S, Whyte P, Ramsey S, Hedman L. WHO guideline on country pharmaceutical pricing policies Geneva: World Health Organization. 2013 [accessed on May 7, 2017]. Available from: http://apps.who.int/iris/bitstream/10665/153920/1/9789241549035_eng.pdf?ua=1.
12. Vogler S, Zimmermann N, Habl C, Piessnegger J, Bucsics A. Discounts and rebates granted to public payers for medicines in European countries. *Southern Medical Review*. 2012;5(1):38–46.
13. Towse A, Garrison LP, Jr. Can't get no satisfaction? Will pay for performance help? Toward an economic framework for understanding performance-based risk-sharing agreements for innovative medical products. *Pharmacoeconomics*. 2010;28(2):93–102.
14. Barinaga G, Chambers MC, El-Othmani MM, Siegrist RB, Saleh KJ. Affordable care organizations and bundled pricing: A new philosophy of care. *The Orthopedic clinics of North America*. 2016;47(4):707–16.
15. Hussey PS, Ridgely MS, Rosenthal MB. The PROMETHEUS bundled payment experiment: Slow start shows problems in implementing new payment models. *Health Affairs (Millwood)*. 2011;30(11):2116–24.
16. Korda H, Eldridge GN. Payment incentives and integrated care delivery: Levers for health system reform and cost containment. *Inquiry*. 2011;48(4):277–87.
17. Robinson JC. Value and payment for oncology in the United States. *Annales Pharmaceutiques Francaises*. 2013;71(5):285–90.
18. Jain SH, Shrank WH. The CMS innovation center: Delivering on the promise of payment and delivery reform. *Journal of General Internal Medicine*. 2014;29(9):1221–3.
19. Licking E, Garfield A, A Roadmap To Strategic Drug Pricing [Internet]. 2016. Available from: http://www.ey.com/Publication/vwLUAssets/ey-in-vivo-a-road-map-to-strategic-drug-prices-subheader/$FILE/ey-in-vivo-a-road-map-to-strategic-drug-prices-subheader.pdf.
20. Greenapple R. Rapid expansion of new oncology care delivery payment models: Results from a payer survey. *American Health & Drug Benefits*. 2013;6(5):249–56.
21. Shih T, Chen LM, Nallamothu BK. Will bundled payments change health care? Examining the evidence thus far in cardiovascular care. *Circulation*. 2015;131(24):2151–8.
22. White C, Chan C, et al. Specialty payment model opportunities and assessment: Oncology simulation report. *Rand Health Quarterly*. 2015;5(1):12.
23. Gavious A, Greenberg D, Hammerman A, Segev E. Impact of a financial risk-sharing scheme on budget-impact estimations: A game-theoretic approach. *European Journal of Health Economics*. 2014;15(5):553–61.
24. Jarosławski S, Toumi M. Market access agreements for pharmaceuticals in Europe: Diversity of approaches and underlying concepts. *BMC Health Services Research*. 2011;11:259.
25. Anastasaki E, Colasante W, Imbeah-Ampiah R. PHP30 can pricing schemes improve market access for innovative high priced drugs? *Value Health*. 2011;14(3):A17.
26. Walzer S, Droeschel D, Shannon R. Which risk share agreements are available and are those applied in global reimbursement decisions? *Value Health*. 2015;18(7):A568.

27. Aggarwal S, Topaloglu H, Messenger M. Novel reimbursement models for cancer drug market access (2010–2013). *Value Health*. 2013;16(3):A153.
28. Morgan SG, Vogler S, Wagner AK. Payers' experiences with confidential pharmaceutical price discounts: A survey of public and statutory health systems in North America, Europe, and Australasia. *Health Policy*. 2017;121(4):354–62.
29. Levy M, Nir RA. The pricing of breakthrough drugs: Theory and policy implications. *PLoS ONE*. 2014;9(11):e113894.
30. Messori A. Application of the price-volume approach in cases of innovative drugs where value-based pricing is inadequate: Description of real experiences in Italy. *Clinical Drug Investigation*. 2016;36(8):599–603.
31. Barry M, Tilson L. Reimbursement of new high cost drugs—Funding the unfundable? *Irish Medical Journal*. 2010;103(5):133.
32. Toumi M, Zard J, Duvillard R, Jommi C. Médicaments innovants et contrats d'accès au marché. *Annales Pharmaceutiques Francaises*. 2013;71(5):302–25.
33. Montazerhodjat V, Weinstock DM, Lo AW. Buying cures versus renting health: Financing health care with consumer loans. *Science Translational Medicine*. 2016;8(327):327ps6.
34. Zettler PJ, Fuse Brown EC. The challenge of paying for cost-effective cures. *American Journal of Managed Care*. 2017;23(1):62–4.
35. Philipson T. Medical breakthroughs and credit markets. *Forbes*. 2014. Available from: https://www.forbes.com/sites/tomasphilipson/2014/07/09/medical-breakthroughs-and-credit-markets/.
36. Gottlieb SCT. Establishing new payment provisions for the high cost of curing disease. Available from: https://www.aei.org/research-products/report/establishing-new-payment-provisions-for-the-high-cost-of-curing-disease/.
37. Carr DR, Bradshaw SE. Gene therapies: The challenge of super-high-cost treatments and how to pay for them. *Regenerative Medicine*. 2016;11(4):381–93.
38. Meghani A, Basu S. A review of innovative international financing mechanisms to address noncommunicable diseases. *Health Affairs (Millwood)*. 2015;34(9):1546–53.
39. Kleinke JD, McGee N. Breaking the bank: Three financing models for addressing the drug innovation cost crisis. *American Health & Drug Benefits*. 2015;8(3):118–26.
40. Beauliere A, Le Maux A, Trehin C, Perez F. Access to antiretroviral treatment in developing countries: Which financing strategies are possible? *Revue d'epidemiologie et de sante publique*. 2010;58(3):171–9.
41. Jack A. Which way now for the cancer drugs fund? *British Medical Journal*. 2014;349:g5524.
42. Mayor S. New "managed access" process for cancer drugs fund to go ahead, NHS England confirms. *British Medical Journal*. 2016;352:i1208.
43. Lu CY. PHP188 Funding the Unfundable: The Australian approach for specialty pharmaceuticals. *Value Health*. 2012;15(7):A322.
44. Marsden G, Towse A, Pearson SD, Dreitlein B, Henshall C. Gene therapy: Understanding the science, assessing the evidence, and paying for value. Available from: https://www.ohe.org/publications/gene-therapy-understanding-science-assessing-evidence-and-paying-value: ICER; 2017.

45. Proach J, Black J, Cromer K, Winters D. Exploring alternative payment methodologies for innovative, expensive treatments: What is needed to trigger a change from the current upfront payment system? *Value Health.* 2016;19(3):A258.
46. Klemp M, Fronsdal KB, Facey K. What principles should govern the use of managed entry agreements? *International Journal of Technology Assess Health Care.* 2011;27(1):77–83.
47. Megerlin F. Conditional pricing for innovative medicines in France: Stop telling about risk-sharing! *Annales Pharmaceutiques Francaises.* 2013;71(5):291–301.
48. Long G, Mortimer R, Sanzenbacher G. Evolving provider payment models and patient access to innovative medical technology. *Journal of Medical Economics.* 2014;17(12):883–93.
49. Malik NN. Pay-for-performance pricing for a breakthrough heart drug: Learnings for cell and gene therapies. *Regenerative Medicine.* 2016;11(3):225–7.
50. Macaulay R. Managed access agreements: A new model pathway for the reimbursement of non-oncology drugs in England approved under European adaptive pathways? *Value Health.* 2016;19(7):A505.
51. Fuller RL, Goldfield N. Paying for on-patent pharmaceuticals: Limit prices and the emerging role of a pay for outcomes approach. *Journal of Ambulatory Care Management.* 2016;39(2):143–9.
52. De Rosa M, Martini N, Ortali M, Esposito I, Roncadori A. An innovative cloud-based platform for implementing performance-based risk-sharing arrangements (PBRSAs) in oncology settings. *Value Health.* 2015;18(3):A273.
53. Dranitsaris G, Dorward K, Owens RC, Schipper H. What is a new drug worth? An innovative model for performance-based pricing. *European Journal of Cancer Care (England).* 2015;24(3):313–20.
54. Garrison LP, Jr., Carlson JJ, et al. Private sector risk-sharing agreements in the United States: Trends, barriers, and prospects. *American Journal of Managed Care.* 2015;21(9):632–40.
55. Launois R, Navarrete LF, Ethgen O, Le Moine JG, Gatsinga R. Health economic value of an innovation: Delimiting the scope and framework of future market entry agreements. *Journal of Market Access & Health Policy.* 2014;2:24988.
56. Brennan TA, Wilson JM. The special case of gene therapy pricing. *Nature Biotechnology.* 2014;32(9):874–6.
57. Messori A, De Rosa M, Pani L. Alternative pricing strategies for cancer drugs. *Jama.* 2015;313(8):857.
58. Touchot N, Flume M. The payers' perspective on gene therapies. *Nature Biotechnology.* 2015;33(9):902–4.
59. Edlin R, Hall P, Wallner K, McCabe C. Sharing risk between payer and provider by leasing health technologies: An affordable and effective reimbursement strategy for innovative technologies? *Value Health.* 2014;17(4):438–44.
60. Stafinski T, McCabe CJ, Menon D. Funding the unfundable: Mechanisms for managing uncertainty in decisions on the introduction of new and innovative technologies into healthcare systems. *Pharmacoeconomics.* 2010;28(2):113–42.

61. Basu A, Subedi P, Kamal-Bahl S. Financing a cure for diabetes in a multi-payer environment. *Value Health*. 2016;19(6):861–8.
62. Avis relatif aux prix de spécialités pharmaceutiques publiés en application de l'article L. 162-16-6 du code de la sécurité sociale. 2019. Available from: https://www.legifrance.gouv.fr/affichTexte.do?cidTexte=JORFTEXT000039668553&categorieLien=id.
63. NHS England Cancer Drugs Fund Team. Appraisal and Funding of Cancer Drugs from July 2016 (including the new Cancer Drugs Fund)—A new deal for patients, taxpayers and industry. 2016 [November 19, 2019]. Available from: https://www.england.nhs.uk/wp-content/uploads/2013/04/cdf-sop.pdf.
64. National institute for Health and Care Excellence (NICE). Cancer drugs fund managed access agreement—Tisagenlecleucel for treating relapsed or refractory B-cell acute lymphoblastic leukaemia in people aged up to 25 years [TA554]. 2018 [November 11, 2019]. Available from: https://www.nice.org.uk/guidance/ta554/resources/managed-access-agreement-december-2018-pdf-6651288397.
65. National Institute for Health and Care Excellence (NICE). Cancer drugs fund managed access agreement—Axicabtagene ciloleucel for treating diffuse large B-cell lymphoma and primary mediastinal B-cell lymphoma after 2 or more systemic therapies [TA559]. 2019 [November 1, 2019]. Available from: https://www.nice.org.uk/guidance/ta559/resources/managed-access-agreement-january-2019-pdf-6660053245.
66. Prasad V, Mailankody S. The UK cancer drugs fund experiment and the US cancer drug cost problem: Bearing the cost of cancer drugs until it is unbearable. *Mayo Clinic Proceedings*. 2016;91(6):707–12.
67. Garau M, Mestre-Ferrandiz J. Access mechanisms for orphan drugs: A comparative study of selected European countries. *OHE Briefing*. 2009;52.
68. Montgomery B. Review of Access to New Medicines. 2016. Available from: https://www.gov.scot/publications/review-access-new-medicines/.
69. Triggle N. Cancer drugs fund "huge waste of money" BBC News. 2017. Available from: http://www.bbc.com/news/health-39711137.
70. Cohen D. Most drugs paid for by £1.27bn cancer drugs fund had no "meaningful benefit." *British Medical Journal*. 2017;357:j2097.

chapter seven

Conclusion

7.1 The long-awaited cell and gene therapies era is finally becoming reality

Gene therapy was first conceptualized in the 1970s [1]. Yet, gene therapies development has been slow, and one of the main setbacks has been the question around the safety of gene therapies. Despite the setbacks, research and development have been ongoing in this field, and after half a century of research, the era of cell and gene therapies is becoming reality [2,3]. Up until September 2019, around 47 cell and gene therapies have been approved globally (nine countries and the EU) and are presented in Table 7.1.

The research and development in this field is growing; according to the Alliance of Regenerative Medicine, around 1,006 cell and gene clinical trials were ongoing worldwide by the end of Q3 2019 [4]. Seventy-eight trials are in phase III of development, expected to reach the market in the coming five years. Products are now knocking at the door and about to flow; regulators and payers need to prepare the roadmap for these promising therapies (Table 7.2).

Table 7.1 Cell and gene therapies approved globally (September 2019)

Trade Name (Active Ingredient)	Therapeutic Indication	Approved Country	Approval Date
Imlygic® (Talimogene laherparepvec)	Local treatment of unresectable cutaneous, subcutaneous, and nodal lesions in patients with melanoma recurrent after initial surgery.	US; EU; Australia	June 28, 2014 December 16, 2015 May 31, 2016
Kymriah® (Tisagenlecleucel)	Patients up to 25 years of age with B-cell precursor acute lymphoblastic leukemia (ALL) that is refractory or in second or later relapse; adults patients with relapsed or refractory large B-cell lymphoma.	US; EU; Canada; Australia; Japan	February 2, 2017 August 22, 2018 September 5, 2018 December 19, 2018 26/03/2019
Yescarta® (Axicabtagene ciloleucel)	Adult patients with relapsed or refractory large B-cell lymphoma after two or more lines of systemic therapy, including diffuse large B-cell lymphoma (DLBCL) not otherwise specified, primary mediastinal large B-cell lymphoma, high grade B-cell lymphoma, and DLBCL arising from follicular lymphoma.	US; EU; Canada	March 31, 2017 August 23, 2018 February 19, 2019
Luxturna® (Voretigene neparvovec-rzyl)	Treatment of patients with confirmed biallelic RPE65 mutation-associated retinal dystrophy.	US; EU	May 16, 2017 November 22, 2018
Provenge® (Autologous Cellular Immunotherapy)	Asymptomatic or minimally symptomatic metastatic castrate-resistant (hormone refractory) prostate cancer.	US; EU	October 30, 2009 September 6, 2013

(*Continued*)

Table 7.1 (Continued) Cell and gene therapies approved globally (September 2019)

Trade Name (Active Ingredient)	Therapeutic Indication	Approved Country	Approval Date
Carticel®	Repair of symptomatic cartilage defects of the femoral condyle (medial, lateral or trochlea)	US	August 22, 1997
MACI®	Repair of single or multiple symptomatic, full-thickness cartilage defects of the knee with or without bone involvement in adults	EU; US	June 27, 2013 December 13, 2016
Zolgensma®	Spinal muscular atrophy (SMA)	US	May 24, 2019
Alofisel® (Darvadstrocel)	Perianal fistulas	EU	March 23, 2018
Spherox®	Articular cartilage defects	EU	July 10, 2017
Zalmoxis®	Patients receiving haploidentical HSCT	EU	August 18, 2016
Strimvelis®	Severe combined immunodeficiency due to adenosine deaminase deficiency (ADA-SCID)	EU	May 26, 2016
Holoclar®	Corneal lesions, with associated corneal (limbal) stem cell deficiency (LSCD), due to ocular burns	EU	February 17, 2015
Chondrocelect®	Used in adults to repair damage to the cartilage in the knee	EU	October 5, 2009
Glybera® (Alipogene tiparvovec)	Adults with lipoprotein lipase deficiency	EU	October 25, 2012
Zynteglo®	Beta thalassemia	EU	May 29, 2019
Allocord® (HPC, Cord Blood)	For use in unrelated donor hematopoietic progenitor cell transplantation procedures	US	May 30, 2013

(Continued)

Table 7.1 (Continued) Cell and gene therapies approved globally (September 2019)

Trade Name (Active Ingredient)	Therapeutic Indication	Approved Country	Approval Date
Azficel-T (Laviv®)	For improvement of the appearance of moderate to severe nasolabial fold wrinkles in adults	US	June 20, 2011
Clevecord® (HPC, Cord Blood)	For use in unrelated donor hematopoietic progenitor cell transplantation procedures	US	June 10, 2015
Gintuit® (Allogeneic Cultured Keratinocytes and Fibroblasts in Bovine Collagen)	For topical (non-submerged) application to a surgically created vascular wound bed in the treatment of mucogingival conditions in adults.	US	March 9, 2012
Hemacord® (HPC-C)	For use in unrelated donor hematopoietic progenitor cell transplantation procedures	US	November 10, 2011
Ducord® (HPC, Cord Blood)	For use in unrelated donor hematopoietic progenitor cell transplantation procedures	US	October 3, 2012
Temcell HS®	Acute GVHD after allogeneic HSCT	Japan	September 18, 2015
Heart Sheet®	Serious heart failure caused by ischemic heart disease	Japan	September 18, 2015
JACC®	Relief of symptoms of traumatic cartilage defects and osteochondritis dissecans (exclude osteoarthritis) for knee joints	Japan	July 27, 2012
Invossa™	Knee osteoarthritis	Korea	July 12, 2017
CureSkin®	Depressed acne scar	Korea	May 11, 2010
CreaVax-RCC®	Metastatic renal cell carcinoma	Korea	May 15, 2007

(*Continued*)

Table 7.1 (Continued) Cell and gene therapies approved globally (September 2019)

Trade Name (Active Ingredient)	Therapeutic Indication	Approved Country	Approval Date
KeraHeal®	Second-degree burns covering more than 30% of TBSA; Third-degree burns covering more than 10% of TBSA	Korea	May 3, 2006
KeraHeal-Allo™	Deep second-degree burns	Korea	October 16, 2015
Holoderm®	Deep second-degree burn; Third-degree burn	Korea	December 10, 2002
Kaloderm®	Deep second-degree burn; Diabetic foot ulcer	Korea	May 21, 2005 (2nd degree burn) June 24, 2010 (Diabetic foot ulcer)
Rosmir®	Improvement of nasojugal groove	Korea	December 28, 2017
Cupistem®	Crohn's fistula	Korea	June 18, 2012
Queencell®	Subcutaneous tissue defect	Korea	March 26, 2010
Cellgram™	Improving heart functions and lowering the MACE through the improvement of Left Ventricular Ejection Fraction (LVEF) in Acute Myocardial Infarction (AMI) patients	Korea	July 1, 2011
Neuronata-R®	Amyotrophic lateral sclerosis	Korea	June 30, 2014
Chondron™	Focal cartilage defect of knee	Korea	January 30, 2001
RMS Ossron™	Focal bone formation	Korea; India	August 26, 2007 April 2017

(Continued)

Table 7.1 (Continued) Cell and gene therapies approved globally (September 2019)

Trade Name (Active Ingredient)	Therapeutic Indication	Approved Country	Approval Date
Immuncell-LC®	Liver cancer (hepatocellular carcinoma)	Korea	August 6, 2007
Cartistem®	Knee articular cartilage defects in patients with osteoarthritis	Korea	January 18, 2012
Chondrocytes-T®	Cartilage lesions associated with the knee, patella, and ankle	Australia	May 26, 2017
Prochymal, also named as TellCell (Remestemcel-L)	Patients NLT 6 months to 17 years of age with acute graft versus host disease (aGvHD), refractory to steroid treatment.	New Zealand; Canada	June 14, 2012; March 6, 2014
Gendicine®	Head and neck squamous cancers	China	October 2003
Oncorine®	Head and neck squamous cancers	China	September 2006
Cartigrow™ (Chondron ACI)	Cartilage defects of the joints	India	April 2017
Apceden™	Four cancer indication: prostate, ovarian, colorectal, and non-small cell lung carcinoma	India	March 2017
Stempeucel®	Critical limb ischemia due to Berger's disease	India	May 2016

Source: Food & Drug Administration (FDA); European Medicines agency (EMA); Pharmaceuticals and Medical Devices Agency (PDMA); Health Canada; Therapeutic Goods Administration (TGA); Ministry of Food and Drug Safety (MFDS); Medicines and Medical Devices Safety Authority (MEDSAFE); Health Sciences Authority (HSA); National Medical Products Administration (NMPA); Central Drugs Standard Control Organization (CDSCO) [5–14].

Table 7.2 Cell and gene therapies clinical trials

	Gene Therapy	Gene Modified Cell Therapy	Cell Therapy
Phase I	115	201	41
Phase II	223	201	147
Phase III	32	16	30
Total	370	418	218

Source: Alliance for Regenerative Medicine Q3 2019 ARM, Quarterly Regenerative Medicine Sector Report, 2019.

7.2 *Regulators prepared for the phenomena*

Regulators have anticipated the novel cell and gene therapies era and established specific regulation for these therapies. These regulations or pathways ensure faster patient access to innovation reaching the market with short-term clinical evidence while the potential for transformative benefits is high, including curative benefits. Therapies are granted approval via these pathways with clinical evidence based on single-arm trials, low sample size, or short follow-up duration. Taking into consideration the potential breakthrough benefit of these therapies, this evidence is considered sufficient to assess the benefit-risk ratio of these products by the regulators.

In order to mitigate the high uncertainty associated with the clinical evidence available at time of launch, several considerations are used by regulators, among which are the risk management plans including up to 15 years of post-marketing exhaustive registry.

It should be noted that the regulations and classifications of cell and gene therapies differ at global level. Different regulatory agencies have adopted different regulations and processes for cell and gene therapies. The lack of harmonization is leading in some cases to divergent decisions. Therefore, an alignment and convergence in regulatory processes are needed to minimize divergent decisions. There were discussions between the FDA and EMA on collaboration for cell and gene therapies regulatory aspects. And it is unclear if the World Health Organization (WHO) will establish a regulatory pre-qualification for cell and gene therapies similar to the ones done for vaccines, biologics, and biosimilars.

7.3 HTA agencies are being conservative, not willing to adapt the current decision frameworks

Due to several reasons described in the previous chapters, cell and gene therapies raise major challenges at the HTA level such as high uncertainty around efficacy and safety. These challenges are not specific to cell and gene therapies; many orphan drugs and biologics have faced this type of challenge. However, in the case of cell and gene therapies, hurdles and challenges are cumulated, at different levels, to a never seen extent, with an unknown magnitude. For example, orphan drugs do not have a single administration with high upfront cost.

HTA agencies' decision framework inadequacy is increasingly singled out in the literature, congresses, blogs, etc. Evidence accumulates about the need to update HTA agencies' framework, and several authors made clear proposals on the methods including the ICER framework for potential cures or single or short term therapies [15]. However, the vast majority of HTA agencies considered their reference case and decision-making framework fully appropriate for cell and gene therapies, refusing the recognition of the specificities of cell and gene therapies compared to other pharmaceuticals.

So far, most HTA agencies are often acting primarily as a lever for reimbursement and/or pricing; they are anxious about the consequences of changing paradigm that may lead to a flow of innovative expensive products, which may afterward constitute a threat for the sustainability of health insurance. Therefore, HTA agencies prefer to stick to an inappropriate model driven by inappropriate requirements and a risk-averse attitude that may delay patient access to a new generation of promising breakthrough therapies. Unlike regulators, HTA agencies have become increasingly conservative bodies and resistant to changes, while regulators developed strong skills to manage risk in order to safely foster patient access to innovation. No HTA framework adaptation is planned in European countries, while ICER in US developed a framework for potential cures.

7.4 Funding remains a serious challenge and little is done today

The high price of cell and gene therapies represents another key challenge. Solutions for affordability challenge are needed to ensure patients' access to innovation while maintaining the health insurance sustainability. Several payment models have been proposed including payment for performance and annuity payment [16]. The latter seems to be the preferred solution by manufacturers while payers resist it. It is still unclear which model payers would be privileged. Payers remain much focused on cost-containment measures.

So far, there is no one-size-fits-all solution; payment models need to be adapted to each country's requirements. However, it should be noted that as detailed in Chapter 6, some payers are starting to adopt new payment models that were uncommon such as payment for performance in Germany.

From the authors' perspective, annuity payment over one year is expected to improve cash flow only, which is not the key issue for most payers. The budget impact still needs to be addressed, and annuity payment will not help to address the substantial budget impact.

The only option for splitting the drug's budget impact over several years would be the depreciation using intangible amortization. Its adoption remains a long route even though it is legitimate.

The healthcoin was considered among the other suggested options to address the change of insurers (payers) in the US on average every 2.5 years [16,17]. However, it is unclear if it would be feasible within the current legal framework in the US.

Special funds have been considered for innovative therapies such as the innovation fund in Italy [18].

7.5 Cell and gene therapies challenge the state-of-the-art business model of pharma industry

The specificities of cell and gene therapies have profoundly disrupted the business model of the pharma industry.

Lower clinical development costs: In some cases, clinical trials with small sample size and short duration are sufficient to achieve marketing authorization for cell and gene therapies. Therefore, the ratio of preclinical and manufacturing investment over clinical development investment in cell and gene therapies may be dramatically higher than the ratio for small molecules or biologics development. The model is leading to lower clinical risk than initially seen with small and large molecules in pharma, and new considerations are required to establish the business model risk and return on investment.

R&D challenges: The nature of the therapies, the single or short-term administration, and curative potential of chronic conditions challenge the current R&D models. The lack of long-term benefits cannot always be compensated by a free additional re-treatment when needed. Indeed, these therapies usually cannot be readministered due to the substantial risk of development of neutralizing antibodies after the first administration. There is a high uncertainty associated with the medium long-term safety of these products, and a lot remains to be discovered. The development of a nonantigenic vector is an option that could be considered by R&D for the future development.

Future competition: The competition is expected to be intense in some disease areas; several new innovative gene therapies targeting the same condition may be approved at the same time leading to an aggressive competition that may decrease the expected business value. This potential risk is foreseen in Huntington's disease and hemophilia, for example.

In addition, the one-time administration leading to cure will change the impact of "first-to-market" therapy. Indeed, the first gene or cell therapy that will reach the market targeting a certain disease will ideally get the chance to treat all prevalent patients. The following cell and gene therapies, depending on the time of their approvals, may compete only for incident patients who usually are significantly smaller target populations. The first-to-market therapy may capture the whole value leaving little or no chance to successive entrants to recover their investment costs.

Furthermore, the competition is expected to be high in this area due to the nature of these products based on "human material replication" that will not allow for a patented drug like small molecules. Only their production processes or technology and use could be patented. The know-how remains a critical asset; however, this suggests a high competition for the same market. For example, there are currently in ongoing development three gene therapies for SMA, five gene therapies for hemophilia, and three gene therapies for Huntington's disease. The know-how transfer is easy to acquire through staff hiring, acquisition of small start-ups, and industrial agreements to access new countries, as was seen in Brazil for vaccines, for example, and is quite common in some countries such as China.

An example of issues around a patent: The company Juno Therapeutics filed a lawsuit in which Juno alleged Kite Pharma scientific collaborators copied research by scientists at Sloan Kettering cancer center to advance Kite's CAR-T work and eventually win approval for Yescarta®. After case assessment, the jury ordered Gilead's Kite Pharma to pay $752 million to Juno Therapeutics and its partners [19].

Attractive area for investment: Although no true breakthrough event has been reached by cell and gene developers through product sales (with the exclusion of merger and acquisition revenue), cell and gene therapies have been attractive assets for investors, especially those willing to risk large amounts of capital on newly formed companies. The return on investment may be through the stock share reward rather than the revenue from these therapies in the medium term. For example, Novartis entered a deal to acquire gene therapy company AveXis for $8.7 billion [20]. On the day Zolgensma® received FDA approval, Novartis stock jumped which constituted a return on the investment done to acquire AveXis [21].

This market is highly attractive for investors from a pricing perspective; gene therapies achieved so far unexpected high prices. However, an intense competition is anticipated, potentially making cell and gene therapies a commodity market, especially after the dissemination of know-how, significant drop of price production, and increased easiness for administration. The future business model of cell and gene therapy manufacturers remains unclear. Despite high uncertainty around these therapies, their development will not be discontinued, especially after the value they will potentially deliver to the society has become obvious and raised unprecedented hopes in the pharmaceutical industry.

7.6 Conclusion

A golden age for gene therapy arrived after half a decade of slow progress. The rapid pace of innovation is bringing revolutionary, potentially transformative therapies to the market. Examples of clinical success are now ample in addition to examples of recent regulatory and reimbursement successes. Around 1,000 cell and gene therapies are in development; because of their versatility, cell and gene therapies are in development in several different disease areas, there are few serious conditions for which gene therapies are not under development.

However, after a fast-growing and rewarding initial phase, the cell and gene therapies environment may significantly change in the future due to the competition, and the progress in the biotechnology field. This market may evolve toward a commodity-based, highly competitive market. The high technical sophistication used for manufacturing may become well established, widely mastered, and available in most countries including some emerging countries due to the lack of product patent and the current era of fast knowledge dissemination. This will ensure a large access and use of these therapies while today cell and gene therapies are considered a challenge for health insurance sustainability.

The duration of the life cycle of cell and gene therapies may be dramatically shortened as seen for medical devices' life cycle, for which companies had faced difficult times to patent their inventions. The need for fast evolution and improvement of the products and associated services may become a critical success factor. However, this should not undermine the breakthrough potential and the innovative characteristics of these therapies.

Finally, despite the setbacks, the advances in the cell and gene therapies field are continuing. The most recent addition to the genome-editing toolbox is the clustered regularly interspaced short palindromic repeats (CRISPR) system, which will potentially be the next revolution to regulate endogenous gene expression [22]. This technique inserts, removes, or replaces specific pieces of the DNA. It may replace the gene replacement therapy currently used to introduce a "healthy" gene to replace the mutated gene. Gene editing represents a highly promising area that raises the potential for curing chronic inherited diseases. With its CRISPR revolution, China may become the number one player based on the high investment in science and technology, the access to a massive local population, and the expertise in this field [23].

References

1. Friedmann T, Roblin R. Gene therapy for human genetic disease? *Science (New York)*. 1972;175(4025):949–55.
2. Hanna E, Remuzat C, Auquier P, Toumi M. Advanced therapy medicinal products: Current and future perspectives. *Journal of Market Access and Health Policy*. 2016;4.
3. Hanna E, Rémuzat C, Auquier P, Toumi M. Gene therapies development: Slow progress and promising prospect. *Journal of Market Access & Health Policy*. 2017;5(1):1265293.
4. ARM. Quarterly Regenerative Medicine Sector Report. 2019 [December 12, 2019]. Available from: https://alliancerm.org/press-release/the-alliance-for-regenerative-medicine-releases-q3-2019-sector-report-highlighting-industry-trends-and-metrics/.
5. Food & Drug Administration (FDA). Available from: https://www.fda.gov/.
6. European Medicines agency (EMA). Available from: https://www.ema.europa.eu/en.
7. Pharmaceuticals and Medical Devices Agency (PDMA). Available from: http://www.pmda.go.jp/.
8. Health Canada. Available from: https://www.canada.ca/en/health-canada.html.
9. Therapeutic Goods Administration (TGA). Available from: https://www.tga.gov.au/.
10. Ministry of Food and Drug Safety (MFDS). Available from: http://www.mfds.go.kr/eng/index.do.
11. Medicines and Medical Devices Safety Authority (MEDSAFE). Available from: https://www.medsafe.govt.nz/.
12. Health Sciences Authority (HSA). Available from: https://www.hsa.gov.sg/content/hsa/en.html.
13. National Medical Products Administration (NMPA). Available from: http://www.nmpa.gov.cn/WS04/CL2042/.

14. Central Drugs Standard Control Organization (CDSCO). Available from: https://cdsco.gov.in/opencms/opencms/en/Home/.
15. ICER. Value Assessment Methods for High-Impact "Single and Short-Term Therapies", November 2019. Available from: https://icer-review.org/material/valuing-a-cure-final-white-paper-and-methods-adaptations/.
16. Hanna E, Toumi M, et al. Funding breakthrough therapies: A systematic review and recommendation. *Health Policy*. 2018;122(3):217–29.
17. Basu A, Subedi P, Kamal-Bahl S. Financing a cure for diabetes in a multi-payer environment. *Value in Health: The Journal of the International Society for Pharmacoeconomics and Outcomes Research*. 2016;19(6):861–8.
18. Flume M, Bardou M, et al. Approaches to manage "affordability" of high budget impact medicines in key EU countries. *Journal of Market Access & Health Policy*. 2018;6(1):1478539.
19. Sagonowsky E. Jury orders Gilead's Kite Pharma to pay $752M for CAR-T patent infringement. 2019. Available from: https://www.fiercepharma.com/pharma/juno-lawsuit-jury-orders-gilead-s-kite-pharma-to-pay-752m-for-car-t-patent-infringement.
20. Novartis successfully completes acquisition of AveXis, Inc. 2018. Available from: https://www.novartis.com/news/media-releases/novartis-successfully-completes-acquisition-avexis-inc.
21. Novartis Gene Therapy Crosses The $2 Million Mark—Will Others Follow? 2019. Available from: https://www.investors.com/news/technology/gene-therapy-zolgensma-approved-spinal-muscular-atrophy-treatment/.
22. Zaman QU, Li C, Cheng H, Hu Q. Genome editing opens a new era of genetic improvement in polyploid crops. *The Crop Journal*. 2019;7(2):141–50.
23. Cohen J, Desai N. With its CRISPR revolution, China becomes a world leader in genome editing. 2019. Available from: https://www.sciencemag.org/news/2019/08/its-crispr-revolution-china-becomes-world-leader-genome-editing.

Index

Note: Page numbers in italic and bold refer to figures and tables, respectively.

A

adrenoleukodystrophy, 25
advanced therapies funding, applicability, 121–122
    annuity-style payment, 115
    bundle payment/episode of care, 106
    intellectual-based payment, 109–110
    international fundraising, 109
    P4P, 110, 115
    price-volume agreements, 107
    rebates/discounts, 107
    reinsurance risk pool, 108
advanced therapy medicinal products (ATMPs), 1, 39, **40–41**
    approved in EU, 7–11
    archetypes, 28, *29*
    classification, 5–6, *6*, 27, **27**
    EU regulation (EC) No 1394/2007 of, 2
    frequency, **21**
    guidelines, 35–36
    hospital exemption, 6–7
    HTA review, EU5, **64–65**
    legislation, *2*
    nonclinical data for, 34, **34**
    overview, 3–5
    regulation, 7, 50
    transitional period timelines, 7
alipogene tiparvovec, 8
allogeneic, 5
    cell therapies, 19
Alofisel®, 10
    England, 76–77
    Germany, 70
annuity payment, pilot test for, 119
annuity-style payment, 115
archetypes, ATMPs, 28, *29*
arthroscopy, 67
Article 8 of Regulation (EC) No 1394/2007, 38
ATMPs, *see* advanced therapy medicinal products (ATMPs)
autologous, 5
    cell therapies, 19
axicabtagene ciloleucel, 10

B

breakthrough therapy designation, 52
bundle payment, 105–106

C

Cancer Drugs Fund (CDF), 75, 93, 121
cardiovascular diseases, 24
CAR-T cells, 119–120
CAR (chimeric antigen receptor) T-cell Therapy, 80–81, 93–94
CAR-T therapies, ICER assessment, 80
CAT, *see* committee for advanced therapies (CAT)
CATP (combined therapy medicinal product), 5
CDF (Cancer Drugs Fund), 75, 93, 121
CED, *see* coverage with evidence development (CED)

cell and gene therapies
approved globally, **128–132**
clinical trials, **133**
funding, applicability, 107
long-awaited era, 127
patients, value for, 23–25
prices for, **22–23**
society, value for, 25–28
specificities, 19–23
state-of-the-art business model, pharma industry, 135–137
in US, 15–16
value drivers, 28
cell replacement therapy, 26
centralized marketing authorization procedure, *38*, 38–39
chimeric antigen receptor (CAR) T-cell Therapy, 80–81, 93–94
CHMP, *see* Committee for Medicinal Products for Human Use (CHMP)
Chondrocelect®, 8
France, 66–67
clinical development costs, 135
CMS coverage, 80–81
combined therapy medicinal product (CATP), 5
committee for advanced therapies (CAT), 2
classifications, *33*
members, 31
scientific recommendation, 32
Committee for Medicinal Products for Human Use (CHMP), 38
compassionate use opinion, 42
Committee on Orphan Medicinal Products (COMP), 39
conditional MA procedure, 42
conditional reimbursement, 91–92
for gene therapies, 93–94
limitation, 92
use, 92
control trial design, 94–95
cost-based/cost-plus pricing, 103–104
cost-containment policies, 102–103
discounts/rebates/expenditure caps/price-volume agreements, 104
ERP, 103
indirect price control, 104
internal reference pricing, 103
VBP, 103
coverage with evidence development (CED), 91–92, 115–116
for CAR-T cells, 93–94
curative therapies, 20, 120–121
curative/transformative effect, 20, 22
cure, 20
current healthcare spending, 101–102

D

darvadstrocel, 10
discounts and rebates, 104, 106–107
dynamic evidence pricing system, 119

E

economic modeling requirements, 96
Electronic Common Technical Document (eCTD), 37, *37*
EMA (European Medicines Agency), 34, 94
episode of care, 105–106
ERP (external reference pricing), 103
EU, *see* European Union (EU)
European Commission (EC), 7
European Medicines Agency (EMA), 34, 94
European Union (EU), 1
ATMPs approved in, 7–11
marketing authorization withdrawal, **85**
reimbursement, 59–61
exceptional circumstances, 8
expenditure caps, 104
external reference pricing (ERP), 103
ex-vivo approach, 3

F

fast-track designation, 51
FDA expedited programs, 51–53
financial advanced therapies agreements, **111–114**
fund-based payment, 108–109
funding models, 104, *105*
features, 116, **117**, *118*

G

G-BA (German Federal Joint Committee), 68–70, 72
GCP (Good Clinical Practice) guidelines, 34
GDP (Gross Domestic Product), 101
gene therapies economic models, **97–98**

gene therapy medicinal product (GTMP), 3
in clinical dossier, 34–35
German Federal Joint Committee (G-BA), 68–70, 72
German health insurance, 119
Glybera®, 8
France, 63, 66
Germany, 68–69
Good Clinical Practice (GCP) guidelines, 34
good manufacturing practice (GMP) compliance, 20
Gross Domestic Product (GDP), 101
GTMP, *see* gene therapy medicinal product (GTMP)

H

healthcare loans/credits, 108
Healthcoin, 116, *118*
health expenditure, *102*
health outcomes-based agreements, 110
annuity payment, 115
CED, 115–116
P4P, 110–115
health-related quality of life (HRQoL), 25
health technology assessment (HTA), 49
agencies' decision framework, 134
benefit-risk assessment, 99, *99*
challenges at, 81–84
commercial challenges, 84
experience in Europe, 63–78
frameworks in Europe, 59–61, **60**
regulators and agencies, 83, *83*
hemophilia, gene therapy for, 26
heterogeneous class, 28
high-cost therapies, funding models, 104
financial agreements, 105–110
Healthcoin, 116
health outcomes-based agreements, 110–116
Holoclar®, 9
England, 75
France, 67
hospital exemption, 6–7
HRQoL (health-related quality of life), 25
HTA, *see* health technology assessment (HTA)
human material replication, 136

I

ICER, *see* incremental cost-effectiveness ratio (ICER); Institute for Clinical and Economic Review (ICER)
Imlygic®, 9
England, 74–75
Germany, 69–70
incremental cost-effectiveness ratio (ICER), 62, 106
initiative, 79, 95
indirect price control, 104
innovation task force (ITF), 48
Institute for Clinical and Economic Review (ICER), 78
intellectual-based payment, 109–110
internal reference pricing, 103
international fundraising, 109
in-vivo approach, 3
Italian model, 92, *93*
Italy, 78
ITF (innovation task force), 48

K

Kymriah®, 11, 24–25
ALL, 68, 75
B cell ALL, 78
CAR-T cell therapies case study, 80–81
DLBCL, 68, 77–78
England, 75
France, 67–68
Germany, 70–72

L

Luxturna®, 11, 24
England, 76
France, 68
Germany, 72–73

M

MAA (marketing authorization application), 33, 36–37
MACI®, 8
managed entry agreement (MEA), 78
market access delays, consequences, 84
marketing authorization
fees, 39, **41**
procedure, *38*, 38–39
timelines, 39
marketing authorization application (MAA), 33, 36–37
MEA (managed entry agreement), 78
micro-, small- and medium-sized enterprises (SMEs), 48
multi-HTA early advice, 96

N

national HTA advice, 96
National Institutes of Health (NIH), 11
national silo funds, 108–109
Net Health Effect (NHE), 62
Netherlands, 92
neuromuscular genetic diseases, 24
new medicines fund (NMF), 121
new payment models, adoption, 119
NHE (Net Health Effect), 62
NICE mock appraisal, 61, **62**
  bridge to HSCT, 62
  curative intent, 63
  hypothesis, 61–62
NIH (National Institutes of Health), 11
NMF (new medicines fund), 121

O

"once and done" concept, 20
oncological diseases, 24–25
orphan drugs, 61

P

P4P (pay-for-performance), 110–115
parallel advice EMA-HTA, 96
Parkinson's disease, 26
pay-for-performance (P4P), 110–115
performance-based agreement, 110
personalized therapies and manufacturing specificities, 19–20
pharmaceutical expenditure, *102*
Pharmaceutical Price Regulation Scheme (PPRS), 104
Pharmacovigilance Risk Assessment Committee (PRAC), 39
post-authorization requirements, 49–50
PPRS (Pharmaceutical Price Regulation Scheme), 104
PRAC (Pharmacovigilance Risk Assessment Committee), 39
preliminary clinical evidence, 52
price caps/volume caps per patients, 107
price-volume agreements, 104, 107
PRIME, 42, **43–47**, *48*
priority review designation, 53
Provenge®, 8
  case study, 86
  England, 73–74
  Germany, 69

R

randomized clinical trials (RCTs), 81
randomized controlled clinical trial, 94
R&D challenges, 135–136
rebates, 104, 106–107, 110
regenerative medicine, 26, 28
Regenerative Medicine Advanced Therapy (RMAT) designation, 1, 11–12, *13*
  for accelerated approval, 53
  application for, 54
  IND for, 52–53
  post-approval requirements, 54–55
  products, **13–15**
regenerative medicine regulatory pathway in EU
  accelerated assessment, 41–42
  ATMPs regulation, 50
  CAT, 31–32
  marketing authorization, 35–41
  post-authorization requirements, 49–50
  premarketing authorization, 32–35
  scientific advice and protocol assistance, 48–49, **49**
regenerative medicine regulatory pathway in US, 51
  FDA expedited programs, 51–53
  post-approval requirements, 54–55
  RMAT designation, 54
  sponsors *versus* CBER review staff, 55, **56**
reinsurance and reinsurance risk pool, 108
risk adjustment, 108
risk-based approach, 32–33
risk corridors method, 108
risk factors, 32
risk management, 50
RMAT designation, *see* Regenerative Medicine Advanced Therapy (RMAT) designation

S

scale-up production, 19
scientific advice and protocol assistance, 48–49, **49**
sCTMP, *see* somatic cell therapy medicinal product (sCTMP)
sipuleucel-T, 8, 73
SMA (spinal muscular atrophy), 24

SMC, 61, 77–78
somatic cell therapy medicinal product (sCTMP), 3–4
  in clinical dossier, 35
Spain, 78
Spherox®, 10
  England, 77
spinal muscular atrophy (SMA), 24
state-of-the-art business model, pharma industry, 135–137
Strimvelis®, 9
  England, 75
substantial manipulation, 4–5
surrogate endpoints, 82

T

21st Century Cures Act, 11–12
talimogene laherparepvec, 9
tisagenlecleucel, 11
tissue-engineered product (TEP), 4–5
tissue therapies, 26
transformative therapies, 22

U

ultra-orphan drugs, 61
uncertainty, 69, 81
  around cost-effectiveness, 82–83
  around long-term efficacy/safety, 82
  around short-term efficacy, 81–82
US reimbursement, 78–79
  advanced therapies in, 80
  CAR-T cell therapies case study, 80–81
  ICER initiative, 79

V

valuation challenges, 84
value-based pricing (VBP), 103
vision disorders, 24
voretigene neparvovec, 11, 24

Y

Yescarta®, 10, 25, 80
  England, 76
  France, 67, 120
  Germany, 72
  Scotland, 77

Z

Zalmoxis®, 9–10
  France, 67
  Germany, 70
Zynteglo®, 11

ed States
blisher Services